RADAR REMOTE SENSING APPLICATIONS IN CHINA

RADAR REMOTE SENSING

APPLICATIONS IN CHINA

Edited by

Guo Huadong

London and New York

Chinese language edition published in 1999 by
Science Press
16 Donghuangchenggen North Street
Beijing 100707

This edition published 2001 by
Taylor & Francis
11 New Fetter Lane, London EC4P 4EE
Simultaneously published in the USA and Canada by
Taylor & Francis Inc,

29 West 35th Street, New York, NY 10001

Taylor & Francis is an imprint of the Taylor & Francis Group

British Library Cataloguing in Publication Data
A catalogue record for this book is available from the British Library
Library of Congress Cataloging in Publication Data
A catalog record for this book has been requested

ISBN 0-415-25676-3

CONTENTS

GEOLOGY

CULTURAL FEATURES AND ARCHAEOLOGY

OCEANOGRAPHY

NATURAL DISASTER

GLOBAL CHANGE

RADAR SCATTERING CHARACTERISTICS AND INFORMATION EXTRACTION

FOREWORD I

Over the last decade, the field of spaceborne imaging radar remote sensing has advanced to the point where many important applications are now possible. With the all-weather, day-night imaging capability, synthetic aperture radar becomes one of the most sophisticated technology for Earth observation and planetary exploration. The deployment of advanced experimental systems, such as the shuttle imaging radar and X band synthetic aperture radar (SIR-C/X-SAR), and operational systems such as ERS-1, JERS-1 and Radarsat have allowed the research and application community to develop the basic understanding and foundation to be able to utilize and interpret radar imaging data to be of direct use for environmental and geophysical problems.

This book touches most of the frontier research disciplines in the radar remote sensing and demonstrates a major contribution to the research and application community to utilize radar systems. Over the last decade, Prof. Guo Huadong and the researchers in the Institute of Remote Sensing Applications of Chinese Academy of Sciences have played a leading role, both in China and internationally, in this field. This book represents an excellent review of application in agriculture, land use, forestry, hydrology, geology, mineral exploration, urbanization, archeology, natural hazards, oceanography and global change seen from scatterometer. In each case a number of illustrative examples of radar images over China are discussed, compared to ground truth and analyzed. In many cases the examples illustrate key advances that have been achieved by IRSA researchers in analyzing radar data.

It includes applications, data analysis, algorithm development, modeling and backscatter behavior analysis of targets such as rice, forest and exposed rock. It illustrate a number of applications such as: 1. Recognition of rice fields with different lifetime, 2.Volume stock estimation of forest with airborne SAR system, 3. Flood monitoring in 1998 with Chinese airborne SAR system, 4. Volcanoes and lava flow detection with SIR-C, 5. Subsurface water revealed by SIR-C, 6. Lithological features underneath vegetation canopy, 7. The twin of Great wall revealed by SIR-C. 8. Internal wave detection, etc. The atlas provides for the first time an extensive and illustrative collection of multi-frequency, multi-polarization, multi-temporal, multi-incidence angle, multi-platform radar images and, in some cases, associated data from other sources over many regions of China. The data was acquired by sensors from international spaceborne and airborne systems from China, European Space Agency, Japan, the former USSR and the USA. All the data was then analyzed and processed at IRSA. This truly reflects on the benefits of international collaborations.

This atlas will be extremely valuable and provide unique value for researchers in all the fields of Earth observation and applications as well as an important element for university students and faculty interested in environmental and geoscientific studies both in China and worldwide. Because of its beautiful layout and color printing it is also of value to non-technical readers with interest in the geography of China.

Dr. Charles Elachi
Academician, National Academy of Engineering of USA
Director, Space and Earth Science Program
Jet Propulsion Laboratory
December, 1998

FOREWORD II

Radar remote sensing has undergone more than thirty years of vigorous development and is now becoming a mature discipline. It has developed from imaging from aircraft platforms to shuttle and satellite imaging systems and from using single band, single polarization to multi-band, multi-polarization systems. By relying on its advantage of nearly all weather imaging capability, its day and night imaging capability, and its longer wavelength penetration capability, radar remote sensing can be used to obtain increasingly more resource information, and its application is becoming widespread in a variety of fields. By the end of the 1990's, radar remote sensing has leaped to a frontier and mainstream position in the development of remote sensing technology and exploitation of remote sensing information in the world. Likewise, radar remote sensing is enjoying abundant support from the Ministry of Science and Technology and has received attention from decision-making organizations. Synthetic Aperture Radar (SAR) research projects have taken precedence for funding applications from China's national space program, 863 High-tech Program, and National Natural Science Foundation Program. The needs of SAR in national defense, resource monitoring, hazard mitigation and other disciplines are increasing daily.

Professor Guo Huadong of the Institute of Remote Sensing Applications, Chinese Academy of Sciences, has been working on radar remote sensing information mechanisms and applications for more than 20 years. He has been responsible for many research projects on microwave remote sensing for earth observations, participated in several large international collaborative programs, and now is the principal scientist of the Expert Group for Information Acquisition and Processing Technology, Hi-Tech Research & Development of China. For the past 20 years he has encouraged young scientists to comprehensively exploit and use radar remote sensing resources. He has prompted them to demonstrate the applications of radar imagery in agriculture, forestry, hydrology, oceanography surveillance, urbanization, and environment as well as geological structure, mineral exploration, oilfield detection, archaeology, etc. He has achieved many breakthroughs, for example, detection of the ancient Great Wall of the Ming and Sui dynasties, classification of subtropical rice fields, recognition of buried rock structures underneath wind blown materials in desert areas, assessment of gold-bearing ore present in altered country rocks, and study of volcanology and geomorphology in the Kunlun Mountains. His research results have been published in international journals and have been cited many times, and results have been positively evaluated at international conferences by our colleagues throughout the world.

I am very pleased to read this book, edited by Professor Guo Huadong. As a leading scientist he has shouldered heavy loads and is hard working year-round, fully occupying every day. Yet he has found time to summarize his many years of research without any reservation to the readers. His devotion reflects his passion for pursuing the scientific spirit. As one of the principal investigators on a series of international radar for earth observation programs, such as SIR-C/X-SAR, Radarsat, ERS-1/2, JERS-1, and GlobeSAR, and as a member of the International SAR Working Group, he has led his colleagues in processing and analyzing radar images acquired from international programs, thereby producing numerous results. This book has selected more than 300 SAR images covering China's land and ocean. It introduces radar remote sensing principles and backscatter characteristics of typical terrain features. By combining the characteristics of geography and environment in China, this book provides many detailed examples and scientific conclusions. Readers can glance over the many radar images from different sensors and locations in China and can use it as a textbook to learn the basics of radar remote sensing. The radar images can also be used to make comparative analyses, to further explore new information, and to resolve technical and environmental problem.

Prof. Chen Shupeng
Academician, Chinese Academy of Sciences
Academician, Third World Academy of Sciences
Academician, International Euro-Asian Academy of Sciences
November, 1998

PREFACE

In the early 1960's, the first synthetic aperture radar (SAR) system was developed, marking a new breakthrough for observing the earth from space by human beings. Since then, radar remote sensing theory, technique, and applications have developed exponentially. In geoscience, bioscience, sustainable development, and other relevant fields, imaging radar has been widely applied. Technology of radar for earth observation evolved from single band, single polarization to multi-band, multi-polarization and has now further evolved to polarimetric SAR, interferometric SAR, spotlight SAR and other new conceptual systems.

Accompanying the worldwide progress of SAR, radar remote sensing in China, which commenced in the 1970's, has been developed in order to accommodate national economic growth and to meet the demands of high technology. Relying on both our own strength and international collaboration, the preliminary experimentation with our radar earth observation system indicates that we have already achieved valuable results. Making use of advanced SAR data from home and abroad, we have made considerable progress in many fields of application.

This book is completed based on this background of development. From a data source point of view, it includes 14 different types of airborne and spaceborne SAR images of China. From the content point of view, it presents an over-all introduction of radar remote sensing principle, technology development, theoretical analysis, and application expansion. Geographically it mainly covers China's land and oceans. This book is composed of ten sections. The first section introduces radar remote sensing principles and technology development. The tenth section discusses radar backscatter characteristics and information extraction methods. The second to ninth sections introduce radar remote sensing applications in eight fields: agriculture, forestry, hydrology, geology, urban and archeology, oceanography, natural disaster, and global changes.

This book serves as both a textbook on radar remote sensing and as an atlas of radar remote sensing. It is rich in text and images and has many distinct characteristics. Some of these are as follows: (1) Advanced SAR data-it comprises data from multiple platforms, multi-bands, multi-polarizations, multiple look angles, multi-temporal, multi-mode, multiple resolution as well as polarimetric SAR and interferometric SAR data. (2) Abundant SAR data-there are five kinds of satellite SAR data, three kinds of space shuttle SAR data, six kinds of airborne SAR data, totaling 340 SAR images. (3) Large time spans-the newest image was acquired in 1998 whereas the oldest was taken in 1979 when the first Chinese airborne SAR image was acquired. (4) Extensive coverage-low resolution SAR images completely cover the Chinese territory whereas middle and high resolution SAR images include 29 provinces, cities, autonomous regions, special administrative regions, and the Bohai Sea, Huanghai Sea, East Sea, and South China Sea (refer to dots on the map of China on the first page of each section which indicate the study areas). (5) New image processing methods-it introduces backscatter models for typical terrain features, advanced algorithms, and information extraction methods as well as data processing methods in the new fields of polarimetric SAR and interferometric SAR. (6) Wide applications-it systematically discusses the contribution of imaging radar to multiple disciplines, demonstrating the important roles that SAR has already played and its bright prospects for future applications in the geosciences, biological sciences, environmental sciences, and other disciplines.

The main co-authors are: Guo Huadong and Shao Yun, Liao Jingjuan, Wang Changlin, Liu Hao; and Wang Cuizhang, Dong Qing, Wang Xiangyun, Wang Chao; and the other authors are Fan Xiangtao, Chi Hongkang, Zhu Liangpu, Lin Qizhong, Shan Xinjian, Chen Zhengli, Lu Xinqiao, Li Junfei, Liu Jianqiang, Zhou Changbao, Wang Shixin, Wang Erhe, Xiao Jianhua, Shao Yiming, Shong Fuming, Sheng Ding, Jing Linjiao, Lou Xiaoguang.

This book is the fruit of many international collaborations. A large number of international collaborative programs of radar for earth observation and organizations, such as NASA/JPL/JSC, DARA/DLR, ISA, CSA, CCRS, ESA, and NASDA, have provided high quality spaceborne SAR data. Without such international collaboration, this book would not have been possible. In particularly, the implementation of SIR-C/X-SAR and GlobeSAR cooperative programs provided a firm foundation for the publication of this book.

In the last six years, the SIR-C/X-SAR project and the China program have benefited greatly from the valuable assistance of M. Baltuck of NASA Headquarters and C. Elachi, D. Evans, T. Farr, J. Plaut, E. Stofen, and E. O. Leary of NASA/JPL. Dr. Elachi, who was a principal scientist of the SIR-A, SIR-B program and an initiator of the SIR-C/X-SAR project has actively promoted bilateral radar remote sensing collaboration. During the GlobeSAR China program from 1993 to 1997, M. St-Pierre of CSA, D. Benmouffok of IDRC, R. Brown and M. D. Iorio of CCRS, and T. Taylor of PCI played important roles. Dr. F. Campbell, a GlobeSAR Coordinator, and B. Brisco, now the President of the Society of Canadian Remote Sensing, have jointly played key roles in our successful completion of the GlobeSAR program in China. I express my appreciation to the above mentioned organizations and individuals and hope our collaboration can go much further.

This book is a result of a long period scientific accumulation. Our research has been funded by the key programs of the National Natural Science Foundation, the Chinese Academy of Sciences and the National 863 High-tech Project 308 subject. During our research, we have been supported by the Geoscience Department and Information Science Department of the State Committee of the Natural Science Foundation; Science and Technology Bureau of Resource and Environment, and the Bureau of Planning and Finance, CAS; Department of High Technology and Industrialization, Ministry of Science and Technology; 863-308 Expert Group; Institute of Electronics, CAS; I extend my sincere thanks to the above mentioned organizations and their officials.

Special thanks are expressed to Academician Chen Shupeng, who from the beginning to the end has guided the writing of this book and devoted much time to this endeavor. Not only has he reviewed the entire book, but he has also written the foreword for the English version. Particular thanks also go to Academician Xu Guanhua for his support approving many of the research programs related to this book and providing academic guidance to our research pursuits. Acknowledgement is also extended to Academician Chen Fangyun and Professor Cheng Jicheng for reviewing the manuscript.

During the editing and publishing period, Professor Yao Suihan and Dr. Liao Jinjuang devoted much creative hard work. Mr. Peng Bin of Science Press gave us great support. Ms. Li Xinfeng provided highly effective work for the publication. Hereby, I express my sincere thanks to all of them. Finally, I would like to express my gratitude to Professor A. J. Lewis of Louisiana State University. Dr. Lewis and his wife Barbara assisted with English translations, primarily grammar and style, from the Chinese version of this book. Without their generous help, the timely publication of the English version of this book would not have been possible.

Although the preparation of this book has spanned many years, as this book is being published we still feel that there is much to be done. We anticipate receiving your views on this book and will be very grateful for your constructive opinions.

Guo Huadong

October, 2000

RADAR REMOTE SENSING APPLICATIONS IN CHINA

Chief-Editor: Guo Huadong

Associate Editors: Shao Yun   Liao Jingjuan

                        Wang Changlin   Liu Hao

Institute of Remote Sensing Applications
Chinese Academy of Sciences, China

Anthony J. Lewis   Barbara A. Lewis
Department of Geography and Anthropology
Louisiana State University, USA

English Editors: Barbara A. Lewis   Anthony J. Lewis

RADAR REMOTE SENSING

Radar remote sensing is an active microwave remote sensing technology, which has near all-weather, day and night imaging capabilities for earth observation and penetration capability for some surface features. The most commonly used electromagnetic spectral ranges; electrons jumping from one stable orbiting shell to another generate visible light. Vibration and rotation of molecules produce infrared and microwave radiation. Microwaves are also generated by fluctuations in electric and magnetic fields. The mechanisms of microwave remote sensing are different from that of optical remote sensing. Radar remote sensing is destined to become one of the most important frontier fields for future earth observations. This is due to a variety of factors, such as the characteristics of imaging radar actively transmitting electromagnetic waves; the sensitivity of radar pulse to surface roughness and dielectric property; the multi-band, multi-polarization scattering characteristics; and the recent development of polarimetric SAR and interferometric SAR imaging methods.

INTRODUCTION TO RADAR REMOTE SENSING

Principles of Radar Remote Sensing

SAR systems are composed basically of a transmitter and modulator, an antenna, a receiver, a data recorder and storage mechanism, and a processor whose output is a SAR image. The output of the transmitter is a train of pulses; a switch directs a pulse to the antenna, which then transmits the pulse to the Earth's surface where it is then scattered. The backscattered signal (radar echo) is detected by the antenna and directed by the switch to the receiver. The receiver's output is then a train of received radar echo pulses that are fed to the correlator.

Spatial resolution is one of the important parameters in a SAR image. It includes both azimuth resolution and range resolution. The time required for the pulse to make the round trip from SAR to a target at slant range R and back is $T=2R/c$, where c is the speed of light and the factor 2 takes into account the round trip of the signal. The resolution r_r along the slant range direction is determined by the pulse width, i.e., $r_r = c\tau/2$. Thus, the resolution in the cross-track direction is given by $r_c = c\tau/(2\sin\theta)$, where θ is the angle of incidence.

Thus, the range resolution is governed by the pulse width and incident angle; range resolution is best at large incident angles or the far range of the image.

The pulse width is determined by the transmitter bandwidth; larger bandwidths produce narrower pulses and thus finer-range resolution. Fine-range resolution is obtainable when the transmitted pulse is chirp modulated (linear FM) and the receiver bandwidth is sufficient to allow matched filtering of the received radar echoes.

In order to achieve fine along-track resolution from imaging radar at orbital altitudes, it is necessary to utilize a Doppler beam-sharpening approach. The basic SÁR technique is to record a series of radar echoes that are received from ground target and that are Doppler shifted due to the motion of the radar and then focus these returns through a special processor and thereby achieve fine along-track resolution. It is important to note that azimuth or along-track resolution is independent of imaging altitude.

Radar Scattering Characteristics

The Radar Backscatter Coefficient

Point targets, such as trucks, aircraft, buildings, corner reflectors, etc., produce a radar echo whose intensity is dependent on the target's radar cross section σ^0, given in units of m^2. Area-extensive distributed targets such as agricultural fields, geologic scenes, or oceanic areas produce a radar echo whose intensity is the radar cross section per square meter. The radar backscatter coefficient σ^0 of an area extensive target is equal to its radar cross section (m^2) divided by its physical area (m^2) in the horizontal plane; therefore, σ^0 is unitless ($m^2\,m^{-2}$). For rough surfaces and /or at near nadir angles of incidence, σ^0 is large. For smooth surfaces and large angles of incidence, σ^0 is small. The backsactter coefficient often exhibits a dynamic range extending over several orders of magnitude, particularly when angular variations are involved.

SAR Image Intensity

After digital processing, the output product of an imaging radar is a digital image composed of a two-dimensional array of pixels. The intensity (also called tone or gray level), or digital number DN (i, j) of the pixel (i, j), is related to the power Pr backscattered from the corresponding ground cell. The received power $P_r(i,j)$ is directly proportional to the radar backscatter coefficient $\sigma^0(i, j)$ of the ground cell, i.e., $P_r(i,j)=k(j)\sigma^0(i, j)$ where $k(j)$ is a range-dependent system factor that incorporates transmitter power, range to target, area of ground cell, and antenna gain function. If $k(j)$ is known on an absolute scale, it is then possible to directly relate DN(i,j) to $\sigma^0(i, j)$ and thereby produce an image whose intensity can be related on an absolute scale to the backsactter coefficient. If only the relative variation of $k(j)$ is known, a relative backsactter image can be generated.

SAR Image Speckle

SAR uses coherent illumination of the scene and coherence of the backscattered signal to achieve high resolution in the azimuth direction. The total signal backscattered from a ground cell is the coherent sum of the signals backscattered from all the scatterers contained in the ground cell. Terrain surfaces usually consist of a large number of randomly distributed scatterers. This randomness is responsible for image speckle, which takes the form of random variability of image tone among pixels corresponding to different cells of a uniform target. Thus, speckle is a sequence of the coherent imaging process employed by SAR systems and is not a result of spatial variability (texture) in the physical or electromagnetic properties of the surface.

Surface and Volume Scattering

The signal backscattered from an area-extensive target may be the result of surface scattering, volume scattering, or both. The relative importance of surface and volume scattering is governed by the surface statistics of the target boundary, the inhomogeneity of the medium underneath the surface, and the penetration depth of the medium. All of these factors are strong functions of or related to the wavelength (λ).

The penetration depth in seawater is only a few millimeters at microwave frequency. Hence, the signal backscattered from the sea contains no volume scattering contribution from the subsurface water, although it may contain volume scattering contributions generated by inhomogeneous layers and other material riding above the water surface. By contrast, backscatter from a vegetation canopy is dominated by volume scattering. Because a vegetation canopy is typically more than 99 percent air by volume, the average relative permittivity of the canopy is only slightly larger than unity. Consequently, the reflection coefficient of the air-canopy boundary is approximately equal to zero, which in turn leads to a negligible surface scattering contribution. The canopy interior, however, is amenable to strong volume scattering because it is very inhomogeneous at the centimeter wavelength scale. If the penetration depth of the canopy is comparable to or smaller than the canopy

slant height (along the direction of observation), contributions from the underlying ground surface may also be present in the backscattered signal. The penetration depth of a medium is determined by both absorption and scattering losses. At centimeter and longer wavelengths, absorption losses exceed scattering losses for most natural media.

Development of Radar Remote Sensing

Airborne SAR was initially the most utilized imaging radar remote sensing system. Developed in the beginning of the 1960's, airborne SAR system was not only a stand-alone airborne system but was also utilized as a test bed for spaceborne SAR systems. SAR has played many important roles over the past thirty years. Typical airborne SAR systems are the CV-580 of the Centre for Canadian Remote Sensing and the AIRSAR system of NASA/JPL. The airborne SAR system in China has undergone 20 years of development, and the newly developed L-band airborne SAR is playing a significant part in many application fields.

Following SAR onboard space shuttle flights in the 1980's, spaceborne radar remote sensing was vigorously pursued with the launch of many SAR satellites, including 1) Almaz SAR of the former USSR, 2) JERS-1 SAR of Japan, 3) ERS-1/2 SAR of ESA, and 4) Radarsat of Canada. Among these, the Canadian Radarsat with its commercial operation ability is characterized by its multi-mode imaging method. The third space shuttle imaging radar (SIR-C/X-SAR), which was developed jointly by the USA, Germany and Italy, was the first SAR system operated simultaneously with multi-frequency and multi-polarization in the earth orbit, and it also had polarimetry and interferometry imaging capabilities. In February 2000, NASA/JPL and DLR jointly conducted a mission called Shuttle Radar Topography Mapping (SRTM), which imaged 80% land surface of the world in a 11-days mission using interferometric SAR technique. This was one of the most challenging and difficult missions in NASA history.

In the future, the European Space Agency will launch an Envisat satellite in 2001 on which an advanced synthetic aperture radar (ASAR) with co-polarization and cross-polarization will be onboard; Canada will launch a Radarsat-2 with full polarimetry in 2003, and Radarsat-3 is planned; Japan is planning to launch an L-band SAR with phase array antenna onboard Advanced Land Observation Satellite (ALOS/PALSAR) in 2003, which has five resolution modes and full polarization capability; and China is planning to launch a SAR satellite in next few years.

A simultaneously multifrequency and multipolarization imaging SAR system can acquire radar responses from terrain features at different bands and co-polarization and cross-polarization information and can accurately detect target characteristics. Polarimetric

SAR can simultaneously receive amplitude and phase information of coherent radar echo, acquiring full polarimetric information including linear, circular and elliptic polarization of terrain features and measuring full scattering matrix of every pixel. Therefore, it can be used in automatic information recognition and surface parameter extraction. Interferometric SAR (INSAR), one of the best techniques for acquiring 3-D information, can be used in numerous ways, such as mapping topography, measuring crustal deformation, or monitoring crop growth status.

SAR Data Used in This Book

This book has used 14 different kinds of SAR data covering the Chinese territory, both land and water, and derived from both airborne and spaceborne SAR systems. Airborne imaging SAR systems include Chinese airborne L-band SAR (L-SAR), CAS/SAR of the Chinese Academy of Sciences, CV-580 SAR of the Centre for Canadian Remote Sensing, AIRSAR of NASA/JPL, Chinese Earth Resource Radar (ERR SAR), and Chinese Real Aperture Radar. Spaceborne SAR data include two categories: space shuttle imaging radar (SIR-A, SIR-B, SIR-C/X-SAR) and satellite SARs (ALMAZ SAR, JERS-1 SAR, ERS-1 SAR, Radarsat SAR) as well as ERS-1 WSC scatterometer data. SIR-C/X-SAR, Radarsat SAR, Chinese L-SAR and Canadian CV-580 SAR are used most frequently in this book.

The editor of this book is a member of the SIR-C/X-SAR science team and a principal investigator (PI) of this project in China. SIR-C/X-SAR data were acquired during two flights in 1994 according to his design for test sites in China. The radar imagery of the Zhaoqing test site in southern China by CV-580 SAR is an important constituent of the GlobeSAR program, which was conducted jointly by the Institute of Remote Sensing Applications, CAS and the Centre for Canadian Remote Sensing. L-SAR data was acquired under the support of the Expert Group for Information Acquisition and Processing Technology, Hi-Tech Research and Development Program in China (863-308). The imaging flight was made by the SAR system, developed by the Institute of Electronics, onboard the remote sensing aircraft of IRSA/CAS. The editor was also a PI of ERS-1/2 SAR, JERS-1 SAR and Radarsat SAR projects. According to agreements, ERS-1/2 SAR and JERS-1 data were provided by ESA and NASDA of Japan respectively; Radarsat data was provided by the Canadian Space Agency's ADRO program, SIR-A and SIR-B data were provided by NASA/JPL. The Institute of Electronics, CAS, developed CAS/SAR system. The PACRIM program of NASA/JPL and the National University of Taiwan provided AIRSAR data.

AIRBORNE SAR SYSTEMS

Chinese L-SAR

The Institute of Electronics of CAS and other research institutions, under the direction of the expert group for Information Acquisition and Processing Technology of the Hi-Tech Research and Development Program of China (863-308), developed L-band airborne SAR system. This system conducted imaging flights during 1997 and 1998 and for the first time acquired Chinese L-band SAR images over land and ocean. In particularly, L-SAR played an important role in monitoring the floods of the Yangtze River in 1998.

L-SAR has two operating modes. Mode "A" is a simulation mode, with radar incident angles varying from 20° to 55°, identical to the proposed Chinese spaceborne SAR. Thus, its imaging geometry and spatial resolution are also the same as that of the Chinese spaceborne SAR.

L-SAR operational modes

Mode	A	B
Flight altitude (km)	6	6
Incident angle(°)	20 ~ 55	66.42 ~ 78.46
Near slant range (km)	6	15
Far slant range (km)	10	30
Slant range swath (km)	4	15
Ground range swath (km)	6	15
Slant range resolution (m)	3	3
Ground range resolution (m)	8.8 ~ 3.7	3.3 ~ 3.1
Azimuth resolution (m)	3	3
Flight speed (km/h)	550	550

Mode "B" is an operational mode; its radar working distance is far, imaging swath is wide, and range resolution is high. This mode is used in airborne SAR experiments.

Chinese Academy Sciences' SAR (CAS/SAR)

The SAR system of the Chinese Academy of Sciences has been under development since 1977. It has undergone three phases: prototype experimental system, single mapping channel system, and multi-channel and multi-polarization system. The SAR system is operated at X-band, pulse repeated frequency is 1000Hz, and pulse peak power is 1 kW. Onboard TY-4 aircraft of the former USSR, the SAR system has a flight altitude of 6000~7000m, flight speed of 450km/h, imaging swath of 9 km, and maximum range distance of 24 km. Because the pulse compression technique was not applied, range resolution was only 180m. However, 30m azimuth resolution was acquired using synthetic aperture techniques. On September 17, 1979, the first CAS/SAR image was acquired in China over the Shanxi Province. In December 1980, the system was improved by increasing transmission peak power to 10 kW, adopting the pulse compression technique, and adding an antenna stabilizing platform and motion compensation circuit. Both range and azimuth resolutions were improved to 15 m.

A single mapping channel SAR system was successfully developed in 1983. In this system, a surface acoustic wave device was used for modulating pulse. A microprocessor was used to control the motion compensation system. Ground speed tracking ability was added. The attitude information of the aircraft was obtained from an inertial navigation system. All these assured high accuracy and high reliability of the SAR system.

In 1987, a multi-channel and multi-polarization SAR system was developed, which was formally named CAS/SAR. The system has the following characteristics: (1) mounting capabilities onboard different kinds of aircraft, working at either high or low altitudes; (2) variability of depression angle of radar beam; (3) utilizing the multi-polarization technique to acquire HH, HV, VH and VV images; (4) using the multiple mapping channel technique to achieve a total imaging swath of 35 km; (5) dual look-direction, looking either to the right or left; and (6) real-time data transmission ability from space to the ground. The following table gives system parameters of CAS/SAR.

CAS/SAR system parameter

System Parameter	Value
Flight altitude (m)	6000 ~ 10000
Ground speed (km/h)	450 ~ 750
Wavelength (cm)	3
Polarization	HH,HV,VV,VH
Depression angle	Variable
Look direction	Right or Left
Resolution (m)	10 × 10 (azimuth × range)
Imaging swath (km)	35
Data recording	Optical

In September 1994, the SAR real-time imaging processor was successfully developed, marking a new level of improvement for CAS/SAR. This processor can process data in real time. The image has a large dynamic range and, therefore, improved image quality.

Canadian CV-580 SAR

Canadian CV-580 SAR onboard Convair 580 aircraft, owned by Canadian Centre for Remote Sensing (CCRS), is a dual frequency, full polarization SAR system with interferometric capability . The system can process data and produce images in real-time. Variable swath width and geometry is possible in three modes: nadir mode, narrow swath mode and wide swath mode, with imaging

swaths of 22 km, 18 km, and 63 km respectively. Each SAR has two receivers and dual polarized antennas so that reception of both like- and cross-polarization is possible. However, in order to record real-time processed data from both channels, the swath width is reduced in half. A fast ferrite switch is being incorporated into the C-band SAR to convert it into a full polarimetric radar, and an additional receiving antenna has been mounted on the side of the Convair to create an interferometric SAR for terrain elevation mapping.

**CV-580 SAR system parameters
in GlobeSAR program**

System Parameter	C-band	X-band
Frequency (GHz)	5.30	9.25
Wavelength (cm)	5.66	3.24
Peak power (kW)	16	6
Polarization	Full Polarization	HH,VV
Interferometry	C-HH	
Resolution (Azimuth × Range)		
Nadir mode (m × m)	6 × 6	6 × 6
Narrow swath mode(m × m)	6 × 6	6 × 6
Wide swath mode (m × m)	20 × 20	20 × 10

In 1993 Canada initiated a GlobeSAR based on CV-580 SAR. As an important member of GlobeSAR program, China has participated in this airborne SAR collaborative remote sensing program along with 11 other countries and conducted processing and analysis of advanced SAR information as well as Radarsat data simulation and potential application studies. On November 20 and 21, 1993, CV-580 SAR was twice flown over the Zhaoqing area of Guangdong Province, during cloudy, rainy weather. The first flight acquired three strips of data, including C-HH, C-VV, X-HH, and X-VV images. The second flight acquired six strips of data, including C-HH, C-HV, C-VV, C-VH data and polarimetric data. During the flights, ground truth data were collected for studies in agriculture, geology, hydrology and forestry. Many field investigations were made in this area in November of the following year, and a great deal of field information was collected.

AIRSAR

AIRSAR is a left looking SAR developed by NASA/JPL onboard a DC-8 aircraft. The system simultaneously operates at three wavelengths: C-band (5.6 cm), L-band (25 cm) and P-band (68 cm). AIRSAR operates in three modes: polarimetric SAR (POLSAR) provides high quality polarimetric data in three frequencies; cross-track interferometric SAR (XTI or TOPSAR) allows precision digital elevation information of the earth's surface to be obtained; and along-track interferometric SAR (ATI) can be used to detect ocean current movement.

POLSAR is fully polarimetric meaning that radar waves are transmitted and received in both horizontal (H) and vertical (V) polarizations for each wavelength. The data products resulting from POLSAR data are in a slant range projection.

In the TOPSAR mode (XTI), data are collected to generate digital elevation models (DEMs) at C-band and L-band. Data products resulting from XTI data are in ground-range projection. Polarimetric data collected in an XTI mode are co-registered to the DEM and are also in ground-range projection. The along-track interferometer (ATI) mode is considered experimental, and the processor for ATI data is still in development.

All TOPSAR interferometers can be operated in single- or dual-baseline modes. For single-baseline operation, signals are transmitted out of one antenna, and the received signals are measured simultaneously through two antennas. In the dual-baseline mode, signals are alternately transmitted out of the antennas at either end of the baseline, while the received signals are measured simultaneously through both antennas.

Radar Modes

• POLSAR: C-, L- and P-band polarimetry (HH, HV, VH and VV polarization combinations).

• ATI: C- and L-band double-baseline along-track interferometry in VV polarization.

• XTI1: C-band single-baseline interferometry in VV polarization; L-and P-band polarimetry.

• XTI1P: C-band double-baseline interferometry in VV polarization; L- and P-band polarimetry.

• XTI2: C- and L-band single-baseline interferometry in VV polarization; P-band polarimetry.

• XTI2P: C-and L-band double-baseline interferometry in VV polarization; P-band polarimetry.

• POLTOP: C-band double-baseline XTI in quad polarization; L-and P-band polarimetry.

Bandwidth

AIRSAR operates at either 20 MHz or 40 MHz bandwidth. An 80 MHz bandwidth is being tested.

• 20 MHz-15 km swath, slant-range resolution 10 x 10 m, DEM resolution 20×20 m.

• 40 MHz -10 km swath width, slant-range resolution 5×5 m, DEM resolution 10×10 m.

• 80 MHz-5 km swath width, slant-range resolution 2×2 m, DEM resolution 10×10 m.

Site Coverage

A data strip for a site is typically 60 km long; adjacent strips can be mosaicked to produce large area images and topographic maps. In regions of high relief where the loss of data from "radar shadow" is common, two data strips of the same site, flown from opposite look direction, can be mosaicked into a single dataset to minimize the data loss.

Earth Resources Radar

The Earth Resources Radar (ERR) is an imaging radar imported from the USA by the National Remote Sensing Center of China. The system is operated at X-band, HH polarization, and has a spatial resolution of 3×3 m; data recording is optical. The SAR system is onboard a Russian made aircraft. For more than ten years, it has imaged many areas within China's territory.

ERR system parameters

System Parameter	Value
Wavelength (cm)	3
Polarization	HH
Look angle (°)	14
Swath (km)	18
Range resolution(m)	3
Azimuth resolution(m)	3
Data recording	Optical
Flight altitude(km)	10

In December 1988, following the request from the Institute of Remote Sensing Application, CAS for the remote sensing team of the National 305 Program, the ERR system made four imaging flights over the Xinjiang test sites.

SPACE SHUTTLE SAR SYSTEMS

Space Shuttle is a space transporting system developed by NASA for travelling between the earth and space and is repeatedly utilized as a multi-function remote sensing platform. In April 1981, the first space shuttle was successfully launched and operated. Shuttle Imaging Radars (SIR-A, SIR-B and SIR-C) were onboard shuttle flights of *Columbia, Challenger*, and *Endeavor* in 1981, 1984 and 1994 respectively and acquired large amounts of SAR data.

SIR-A

SIR-A was an L-band, HH polarization SAR with look angle of 47°and azimuth and range resolutions of 40 m × 40 m. Radar beams were transmitted through a synthetic aperture antenna towards the ground. An optical recorder then recorded the radar echoes. Radar images were then produced after processing the recorded signals with an optical correlator. The operation parameters of SIR-A are given below.

In November 1981, SIR-A imaging flights for large portions of the earth were acquired. Three data strips

SIR-A operation parameters

System Parameter	Value
Frequency (GHz)	1.278
Wavelength (cm)	23.5 (L-band)
Polarization	HH
Look angle(°)	47
Imaging swath (km)	50
Range resolution (m)	40
Azimuth resolution (m)	40
Number of looks	6
Data recording	Optical
Flight altitude (km)	260
Launch date	Nov. 12, 1981
Flight duration (day)	2.4
Mission spacecraft	*Columbia*

were collected over China: 1) the northern two strips passed over the middle section of the Huabei area, northern Ganshu Province and the middle part of Xinjiang Uygur Autonomous Region, and 2) the southern strip passed over Hainan Island, Yunnan Province and Tibet.

SIR-B

With the success of SIR-A, the radar was upgraded significantly. SIR-B, the follow-up to SIR-A, was launched into space in October 1984. In comparison to SIR-A, the most significant change in SIR-B was its flexibility of imaging geometry, particularly the effect of incident angle on radar backscatter. The antenna panels were mounted on a new support structure, which

SIR-B operation parameters

System Parameter	Value
Frequency (GHz)	1.282
Wavelength (cm)	23.4 (L-band)
Polarization	HH
Look angle (°)	15 ~ 60
Imaging swath (km)	20 ~ 40
Range resolution (m)	58 ~ 17
Azimuth resolution (m)	25
Number of looks	4
Data processing	Digital and optical
Flight altitude (km)	354, 257, 224
Inclination (°)	57
Launch date	Oct. 5, 1984
Flight duration	8.3
Mission spacecraft	*Challenger*

allowed them to be mechanically tilted to any look angle between 15° and 60°. With the variation of look angle, the range resolution, imaging swath, signal to noise ratio, and dynamic range of SIR-B were varied accordingly. The range resolution was improved from 40m for SIR-A to up to 20m for SIR-B. The data recording method was changed from optical recording to both optical and digital recording. There was one data strip collected over Xining City in the middle part of China.

SIR-C/X-SAR

Spaceborne Imaging Radar C/X-band Synthetic Aperture Radar (SIR-C/X-SAR) was flown aboard two space shuttle flights in April and October 1994, conducting two ten day missions for earth observation from space. The SIR-C/X-SAR missions returned 107 terabits (10^{12}) of data for more than 300 test sites in the world and were very successful.

SIR-C/X-SAR was a large international spaceborne radar program for earth observations. NASA, the German Space Agency, and the Italian Space Agency jointly developed its hardware. In addition, scientists from 13 other countries, including Australia, Canada, China, United Kingdom, France, and Japan, joined this international collaborative program. The program was composed of 52 research teams, including the Chinese team.

SIR-C/X-SAR has three distinctive characteristics. It was the first spaceborne radar to operate simultaneously at several frequencies; it was the first spaceborne radar to operate simultaneously at different polarizations; and it demonstrated SAR interferometry technology for the first time from the space shuttle platform. This system, to a large extent, has made use of frequency, polarization, phase and amplitude information for electromagnetic waves and provides an effective means for recognizing terrain features.

As a part of the SIR-C/X-SAR program, the investigation team in China conducted multi-discipline research work. In synchronization with shuttle radar flight, airborne SAR imaging flights and field observations were conducted. Making use of the advanced SAR data facilitated detection and recognition of the terrain surface and subsurface as well as man-made objects.

SIR-C/X-SAR is operated at L-, C- and X-band and can image the same area at the same time. At L- and C-band, there are four polarization combinations, HH, HV, VH and VV. SIR-C can also provide polarimetric SAR data. X-band is VV polarization. Look angle ranges from 15° and 60° and is steerable on-board the Shuttle. Some selected data were transmitted through a Tracking and Data Relay Satellite (TDRS) to the ground for real-time analysis.

SIR-C/X-SAR has developed and used four

advanced techniques: interferometric SAR (INSAR), ScanSAR, calibration and SAR data real-time processing. INSAR data, including Chinese test sites, was collected during the last three days of the SIR-C second flight. In order to meet the needs of large-scale research for resource and environment, ScanSAR mode with an imaging swath up to 200 km was applied for two flights of SIR-C/X-SAR. SAR data calibration can provide backscatter coefficients of terrain features, which are useful in quantitative remote sensing analysis. The successful development of the SAR real-time ground processor (a high-speed processor with data rate 45Mbit/s) allows a continuous output of SAR image strips. Data from 7-km image strips can be processed in a second.

In 1990 we began to select the test sites in China for the SIR-C/X-SAR program. After the system parameters and imaging capabilities of SIR-C/X-SAR had been decided, we selected six test sites in China for our study, including areas in Xinjiang, Inner Mongolia, Huabei, Guangdong, Hubei, and Taiwan. These SIR-C/X-SAR data were used in studies of geology, forestry, oceanography, SAR penetration, etc. During the flights in April and October, 1994, SIR-C/X-SAR data were acquired for all test sites in China according to our designed imaging requirements. Depending on the time of SIR-C/X-SAR operation, the imaged length for each test site varied; some were in several hundred kilometers, and some exceeded 2000 km. For example, the Guangdong test site includes Hainan, Hongkong, and other areas; Huabei test site includes Beijing, Hebei, Shangdong, etc. A total of 90 minutes of data was

SIR-C/X-SAR system parameters

System parameter	L-band	C-band	X-band
Flight altitude (km)		225	
Wavelength (cm)	23.5	5.8	3.1
Polarization		HH, HV, VH, VV	VV
Azimuth resolution (m)		30 × 30	
Range resolution (m)		13 × 26	10 × 20
Imaging swath (km)		15 ~ 90	15 ~ 60
Look angle (°)		20 ~ 55	
Band width(MHz)		10, 20	
Pusle length (μ s)		33.17, 8.5	40
Data rate (Mbit/s)		90	45
Acquisition time (h)		>100	

Note: The parameters of L-, and X-bands are the same with C-band except mentioned in the table.

SIR-C/X-SAR operation modes

Mode	Channel A	Channel B	Channel C	Channel D	Imaging data takes
1	-	-	-	-	Only for X-SAR
2	L-HH	L-HH	L-HV	L-HV	Two data takes
3	L-HV	L-HV	L-VV	L-VV	Two data takes
4	C-HH	C-HH	C-HV	C-HV	Two data takes
5	C-HH	C-HH	L-VV	L-VV	Two data takes
6	C-HH	C-HH	C-VV	C-VV	Two data takes
7	L-HH	L-HH	C-VV	C-VV	Two data takes
8	L-HH	L-HH	C-HH	C-HH	Two data takes
9	L-VV	L-VV	C-VV	C-VV	Two data takes
10	L-VH	L-VH	C-HV	C-HV	Two data takes
11	L-HH	L-HV	C-HH	C-HV	One data take
12	L-VH	L-VV	C-VH	C-VV	One data take
13	L-HH	L-VV	C-HH	C-VV	One data take
14	L-HH	L-VH	L-HV	L-VV	One data take
15	C-HH	C-VH	C-HV	C-VV	One data take
16	L-HH,VH	L-HV,VV	C-HH,VH	C-HV,VV	One data take
17	L-HH	L-HH	L-VV	L-VV	One data take
18	C-HH	C-HH	C-VV	C-VV	One data take
19	L-HH	L-HH	C-HH	C-HH	One data take
20	L-VV	L-VV	C-VV	C-VV	One data take
21	L-VH	L-VV	C-VV	C-VV	One data take,Interferometry
22	C-VV	C-VV	C-VV	C-VV	One data take,Interferometry
23	L-VH,HH	L-VV,HV	C-VH,HH	C-VV,HV	One data take

collected for China during these two flights of SIR-C/X-SAR. In addition, the astronauts of Shuttle Endeavor took many photos for these test sites using a long focal length camera.

SATELLITE SAR SYSTEMS

Canadian Radarsat

Radarsat is Canada's first earth observation satellite; it was constructed and launched under the aegis of the Canadian Space Agency in cooperation with the U.S., the provincial governments, and private sectors. The primary objective of the Radarsat mission is to provide application benefits related to offshore oil and gas exploration, ocean fishing, shipping, agriculture, geology, forestry, and land use. It was launched in November, 1995.

Radarsat is carrying a C-band SAR, operated in HH polarization and five imaging modes, i.e. fine resolution, standard, wide (1, 2), ScanSAR (narrow, wide), and extended (high, low). In comparison with other SAR systems, Radarsat has the following characteristics: (1) imaging capability at swath 45, 75, 100, 150, 300 and 500 km; (2) resolution selectable, depending on the applications, with three bandwidths of 11.6, 17.3, 30.0 MHz available; and (3) powerful data processing ability.

The multi-function ability of Radarsat has enhanced its imaging capability. It can cover from 73°N to the entire area comprising the Arctic in 24 hours, all of Canada and northern Europe in three days, and the entire globe in 24 days.

European Remote Sensing Satellite SAR (ERS-1)

The ERS-1 satellite was launched on July 16, 1991 to a near-polar, sun-synchronous orbit. The orbit has an altitude of 790 km and an inclination of 98.5°. The ERS-1 SAR is a part of the Advanced Microwave Information (AMI) package, a unit utilizing shared electronics for the SAR and a scatterometer which can be operated in either a wind or wave mode. As a result, the operation of the SAR and scatterometer are mutually exclusive. The C-band SAR has VV polarization and an incident angle of 23°. The swath width is 100 km wide with 30 m resolution at 4 looks.

ERS-1 SAR technical parameters

Parameter	Imaging Mode SAR	Wave Mode SAR
Band	C	C
Transmitting pulse width (s)	37.12	7.12
Peak power loss (W)	1135	284
Antenna size (m × m)	10 × 1	10 × 1
Polarization	VV	VV
Incident angle(°)	24	24
Data rate (Mbit/s)	102	0.345
Swath (km)	99.6	5 × 5
Spatial resolution (m)	26	26
Radiometric resolution (dB)	1.79	1.70
Satellite altitude (km)	790	790

Radarsat SAR imaging modes

Imaging Modes	Resolution (m) (range × azimuth)	Looks	Swath (km)	Incident Angle (°)
Radarsat				
Standard	25 × 28	4	100	20~49
Wide (1)	(48~30) × 28	4	165	20~31
Wide (2)	(32~25) × 28	4	150	31~40
Fine Resolution	(11~9) × 9	1	45	37~48
ScanSAR (N)	50 × 50	2~4	305	20~40
ScanSAR (W)	100 × 100	4~8	510	20~50
Extended (H)	(22~19) × 28	4	75	42~60
Extended (L)	(63~28) × 28	4	170	10~25

Japanese Earth Resources Satellite (JERS-1)

In 1992, Japan launched the first Japanese Earth Resources Satellite (JERS-1), aimed at establishing an integrated system of optical and microwave sensors for examining terrestrial resources and the environment, focusing on geological and topographic surveys. The SAR is an L-band, HH polarized system with a 15 MHz bandwidth. Data was acquired at an incident angle of 35°. The swath width was 75 km with a resolution of 18 m (3 look) in the range direction and 18 m in the azimuth direction. The JERS-1 satellite was placed in a sun-synchronous orbit 568 km above the earth, giving a 44-day westward repeat cycle and thus allowing access to almost the entire globe with a fixed SAR look angle. Japan Earth Observation Center delivered JERS-1 SAR scenes to users either on a tape of 75 km × 75 km or as a 24 cm black and white image.

JERS-1 SAR operation parameters

System Parameter	Value
Band	L
Polarization	HH
Incident angle (°)	35
Resolution (m)	18 × 18
Swath (km)	75
Satellite altitude (km)	568
Orbit inclination (°)	97.7
Bandwidth (MHz)	50
Data rate (Mbit/s)	74
Launch time	February 1992

Russian Almaz SAR

On March 31, 1991, the former Soviet Union launched a spaceborne SAR called Almaz-1. The Almaz-1 SAR is a single frequency S-band (10 cm wavelength), single polarization imaging system with a spatial resolution of 15 to 30 m, depending on the incident angle, and a swath width of 20 km. Incident angles are selectable between 30° and 60°. This was the first spaceborne SAR launched in the 1990s. However, the Almaz-1 stopped operating after ten months in service. Nevertheless, Almaz SAR provided a large amount of data between latitudes 78°N and 78°S.

ALMAZ SAR operation parameters

System Parameter	Value
Orbital altitude(km)	300
Incident angle (°)	30~60
Band (wavelength, cm)	10
Polarization	HH
Range/azimuth resolution (m)	15~30
Radiometric resolution (dB)	3~5
Pulse power (kW)	190
Antenna size (m)	15 × 1.5
Data record method	Tape aboard

ERS-1 Windscatterometer

Long term and global measurements of the normalized radar cross section (NRCS) of the Earth's surface became available with the launch of the Windscatterometer (AMI-Wind) aboard the First European Remote Sensing Satellite ERS-1 of the European Space Agency in July 1991. The windscatterometer operates at C-band (5.3 GHz) with VV polarization. Its three antennae look 45 degrees forward (fore), sideways (mid), and 45 degrees backward (aft) with respect to the satellite flight direction. The three antennae illuminate a 500km wide swath, in a quasi-simultaneous mode, as the satellite moves along its orbit. Across the swath, local incident angles range from about 18° to 47° for the midbeam and from 25°to 59° for the two other antennae. Data provided by the ESA are given with a pixel size of 25km, but the actual resolution is about 50km.

Airborne SAR system of the Chinese Academy of Sciences (CAS/SAR)

Shuttle imaging radar C/X-band SAR system

Canadian CV-580 SAR system

Canadian Radarsat SAR system

American AIRSAR system

European Remote Sensing Satellite (ERS-1)

AGRICULTURE

Farming activities are the most important living activities of human beings, and are essential for the existence and development of a healthy society. The estimation of crop yield is a topic of global interest, and the efficient management of agricultural land resources is strongly related to social and economic sustainable development, especially in China. It is well known that China is the largest country in population, which has strong influence on agricultural activities and food supply. Crop yield has increased every year over the last five decades in China. However, as population has increased, economic and industrial development have also taken place, resulting in a decrease in the quantity and quality of cultivated land in China. Sufficient food supply for China's current population of 1.2 billion people is a serious concern facing China, a concern that will intensify as the population continues to grow. Therefore, it is important to find an efficient way to face this dilemma. Additionally, as China is entering the international grain market, precise information about crop production and land resources becomes more and more valuable.

Remote sensing technology provides dynamic, accurate, and timely information allowing for monitoring agricultural activities and land resources, and forecasting crop yield. This chapter selects test sites located in the Pearl River Delta, Yangtze River Delta, North China Plain, and irrigated lands in desert environments to demonstrate the capability and potential of radar remote sensing for crop discrimination, growth status monitoring, and land cover mapping.

CROP MONITORING AND LAND COVER MAPPING IN SOUTH OF CHINA

Optical remote sensing technology has been successfully applied to land cover mapping and crop yield estimation in the north of China. However rice crop yield estimation and land cover mapping in the south of China remains a challenge to the remote sensing community because of the poor availability of a large amount of revisit data in this cloudy, rainy climate. With its all-weather observing capability, synthetic aperture radar (SAR) not only serves as a complimentary remote sensing data source to optical remote sensing but becomes a major source of fine to medium resolution data in cloudy humid environments. This section mainly presents the results of operational radar remote sensing for rice monitoring and yield estimation as determined from the "SAR Technology for Rice Growth Status Monitoring and Land Cover Mapping" project under the National High Technology Program (863-308) funded by the Ministry of Sciences and Technology. Most images in this section were acquired by the Canadian Radarsat centered at south china around 23°2' N and 112° 7' E . A few images were also collected in the Yangtze River Delta.

Rice Monitoring and Land Use Mapping in Zhaoqing

The Zhaoqing test site is located in a humid subtropical environment. Zhaoqing is economically a rapidly developing region with a well-developed transportation system and rich natural resource base. The total annual sun illumination is 1900 hours, the average annual temperature is between 21.5 ~ 21.9°C, the annual accumulate temperature is 7600 °C for $\geqslant 10$°C, and the annual rainfall is 1876 mm. The combination of low temperatures and heavy rainfall indicates the hazards that tropical cyclones pose to crops. Sufficient heat and seasonal water supply make Zhaoqing an ideal region for crop production. These favorable conditions result in complex land cover patterns that include paddy fields, orchards, woodlands, fish farming, water plants, tropical vegetation, and other land uses.

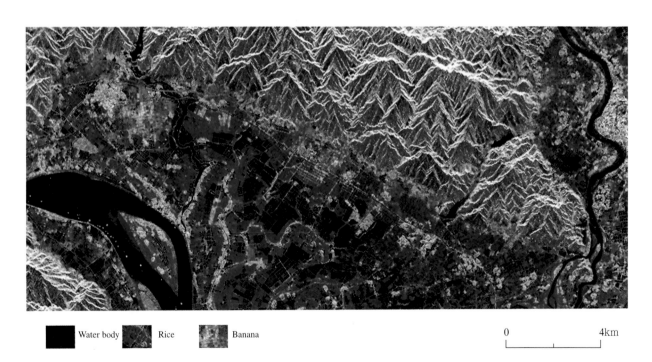

Water body　　Rice　　Banana　　　　0　　　4km

Color composite SIR-C/X-SAR image of Zhaoqing site (R: L-HV, G: X-VV, B: L-HH).

Early Stage Rice Recognition

The multi-frequency, multi-polarization SIR-C/X-SAR image of Zhaoqing was acquired on April 18, 1994. In the false color composite SIR-C/X-SAR image, acquired two weeks after transplanting, rice is green, water bodies are black, banana fields are light blue, *Euryale ferox* (an aquatic plant related to lotus) is dark green, residential areas located on the alluvial plain are pink, and forests on the hills are pink as well. The single parameter (black and white) image shows that the backscatter intensity of rice decreases as wavelength increases. *Euryale ferox* can be discriminated from water only on the X-VV image, which illustrates that VV polarization is more sensitive than HH polarization to a slightly rough surface.

Single parameter SIR-C/X-SAR image
(top to bottom: L-HH, L-HV, C-HH, C-HV, X-VV).

Rice Growth Status Monitoring and Land Cover Mapping

Multi-temporal, multi-mode Radarsat images have great potential for rice monitoring and land cover mapping. A number of targets have been identified on multi-temporal false color Radarsat imagery acquired on April 25, August 23 and November 27, 1996. Sugar cane is in yellowish green in the right corner; banana and residential areas are in white at the center, *Euryale ferox* is in dark green, medium mature rice is in blue, and late mature rice is in magenta. This multi-temporal false color image distinguishes two kinds of sugar cane based on the difference in their backscatter coefficients on the image of April 25. However, the late mature rice and flooded point bars and other river lowlands have the same backscatter coefficients in the image of August 23 due to flooding in the rainy season. These two

Standard mode Radarsat image of Zhaoqing site (1996).

0   4km

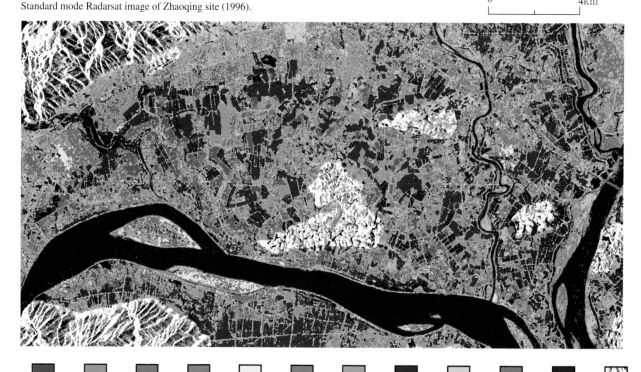

| Medium mature rice 1 | Late mature rice | Early mature rice | Medium mature rice 2 | Banana | Sugar cane 1 | Sugar cane 2 | *Euryale ferox* | Grassland | Residential area | Water body | Forest |

Land cover map of Zhaoqing site (based on standard mode Radarsat image of 1996).

14

Multi-temporal fine-mode Radarsat image (1996) (R: Aug. 4, G: Nov. 8, B: Sept. 21).

Medium mature rice

Late mature rice

Late transplanted rice

Early season rice

Late season rice

Euryale ferox

Banana

Grassland

Lotus

Residential area

Water body

Forest

0   2km

Crop map (based on fine-mode Radarsat image of 1996).

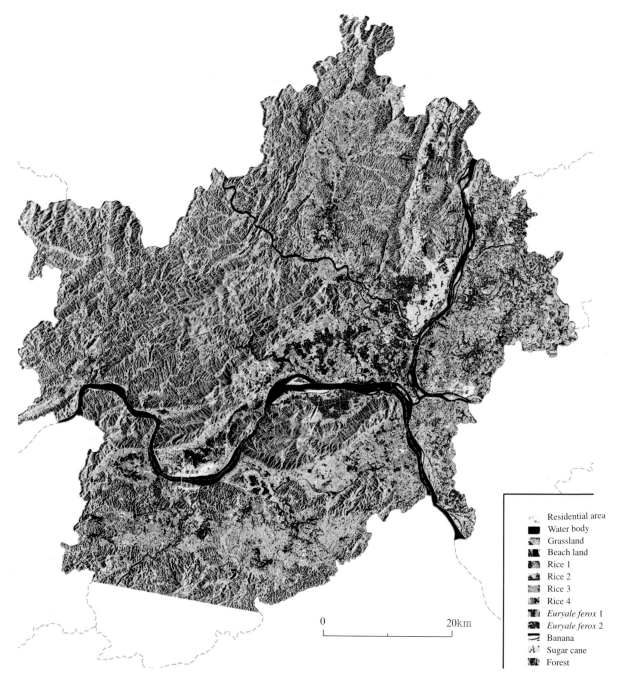

	Residential area
	Water body
	Grassland
	Beach land
	Rice 1
	Rice 2
	Rice 3
	Rice 4
	Euryale ferox 1
	Euryale ferox 2
	Banana
	Sugar cane
	Forest

0   20km

Multi-temporal false color standard mode Radarsat image map of Zhaoqing site (1997)
(R: July 22, G: April 25+May 19, B: June 4+June 12+June 28).

examples indicate the importance of the time of data acquisition. A land cover map was produced based on classification results of the multi-temporal Radarsat image with a neural net classifier. The mountainous areas were manually delineated. Eleven types of targets were discriminated.

The multi-temporal Radarsat image successfully identified five classes of rice in 1996 and four classes of rice in 1997. The differences were due to the type of rice planted, the stage of growth, and farming practices. In the fine-mode Radarsat image, medium mature rice

(MM) is blue, late mature rice (LM) is light pink, and late-transplanted rice (LTP), transplanted 25 days later, is red. Early season (ES) and late season (LS) represent the single rice crop in spring and autumn respectively. The land cover map was produced based on the fine-mode Radarsat image.

Seven scenes of Radarsat images were acquired in 1997. With an understanding that radar backscatter from rice is a function of time, test sites in Sihui, Gaoyao and Sanshui counties, and Dinghu and Duanzhou administration regions, totaling 5000 km^2, were selected

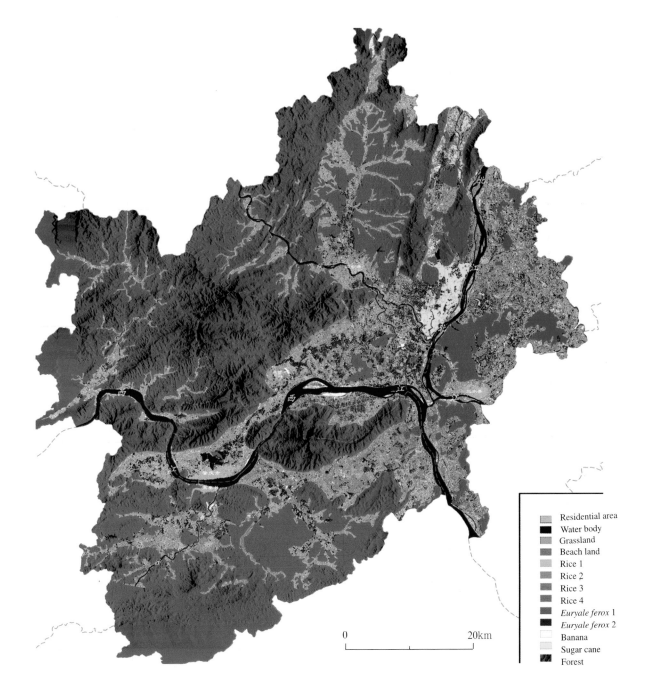

Land cover map of the Zhaoqing site.

Residential area
Water body
Grassland
Beach land
Rice 1
Rice 2
Rice 3
Rice 4
Euryale ferox 1
Euryale ferox 2
Banana
Sugar cane
Forest

0   20km

to produce the multi-temporal false color Radarsat image, the land cover map, the crop distribution map, and the rice distribution map. Twelve types of target were classified using a neural net classifier. According to their life span, rice crops are discriminated as (1) medium mature rice, (2) late mature rice, (3) medium-late mature rice, and (4) early mature rice. It is recognized that *Euryale ferox* (1) was planted earlier than *Euryale ferox* (2) and is therefore at a more luxuriant stage of growth. Precise calculation of the acreage of planted rice, identification of rice with different life spans, and the estimated yield of rice at this test site are discerned.

Field photo of rice taken at times of transplanting, seedling developing, and ear differentiating

17

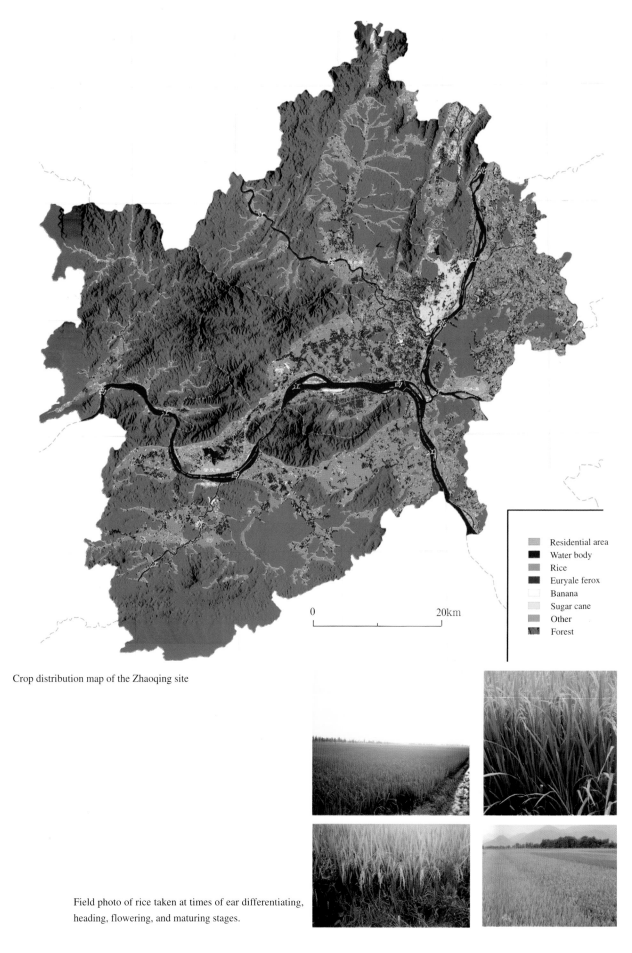

	Residential area
	Water body
	Rice
	Euryale ferox
	Banana
	Sugar cane
	Other
	Forest

0   20km

Crop distribution map of the Zhaoqing site

Field photo of rice taken at times of ear differentiating, heading, flowering, and maturing stages.

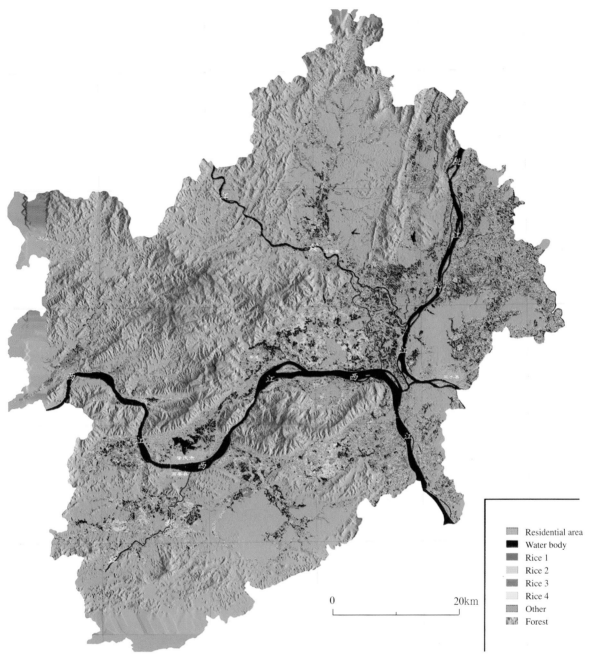

Rice distribution map of the Zhaoqing site.

	Residential area
	Water body
	Rice 1
	Rice 2
	Rice 3
	Rice 4
	Other
	Forest

0   20km

Field photo of spring rice at harvesting and autumn rice at transplanting stage.

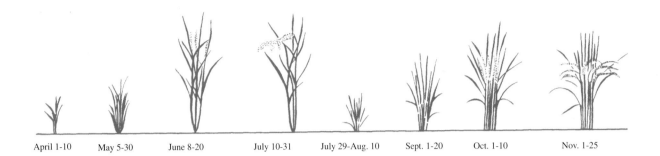

| April 1-10 | May 5-30 | June 8-20 | July 10-31 | July 29-Aug. 10 | Sept. 1-20 | Oct. 1-10 | Nov. 1-25 |

Rice calendar for double cropping rice (spring rice and autumn rice).

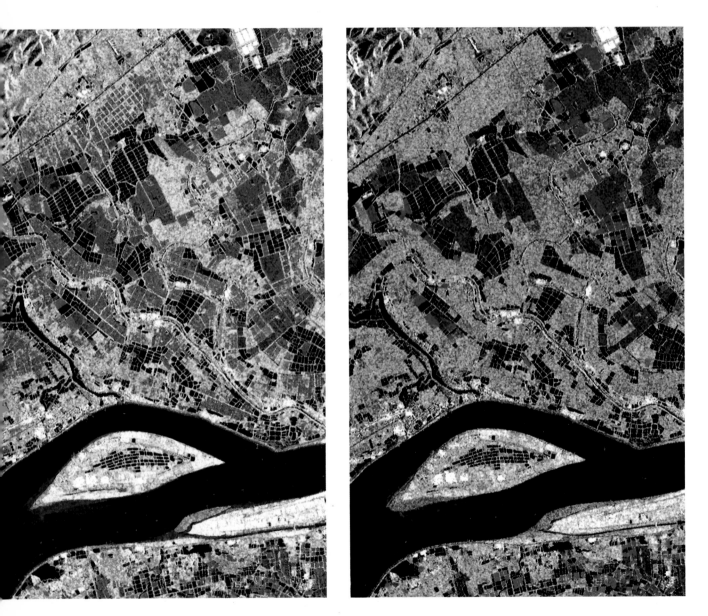

Multi-temporal fine-mode Radarsat image (Left R:Nov. 1996, G:Aug. 1996, B:Sept.1996; Right R:Dec.1996, G:June 1996, B:Sept.1996)

Land Cover Mapping

This false color multi-frequency, multi-polarization airborne radar image was acquired by a GlobeSAR mission conducted from 8 to 12 pm on November 20,

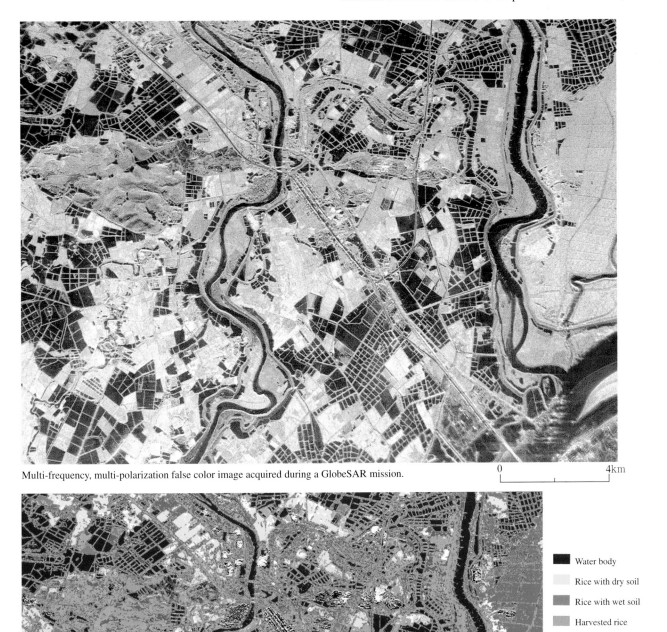

Multi-frequency, multi-polarization false color image acquired during a GlobeSAR mission.

0   4km

Land cover map based on GlobeSAR image.

- Water body
- Rice with dry soil
- Rice with wet soil
- Harvested rice
- Rice with water
- Sugar cane
- Sugar cane (20 days early)
- Banana
- Grassland
- *Euryale ferox*
- Trees
- Shadow or bare soil
- Residential area
- Other crops

1993. The image contains a considerable amount of information. The variation of the color and tone on the left corner of the image represents harvested rice or rice fields with different water contents under the rice canopy. Euryale ferox is dark blue due to its strong return on VV polarization image. The C-HH, C-HV, C-VV, X-HH, X-VV images were co-registered, then classified with a neural net classifier. Eleven types of targets were discriminated. This study demonstrates the capability of multi-frequency, multi-polarization radar for target recognition.

The high-resolution airborne L-SAR data of Zhaoqing was acquired in September 1997. This image is part of an orthographic digital mosaic. Medium mature rice, late mature rice, banana, and sugar cane are delineated on the L-SAR image. Because it is an L band image the *Euryale ferox* cannot be separated from water.

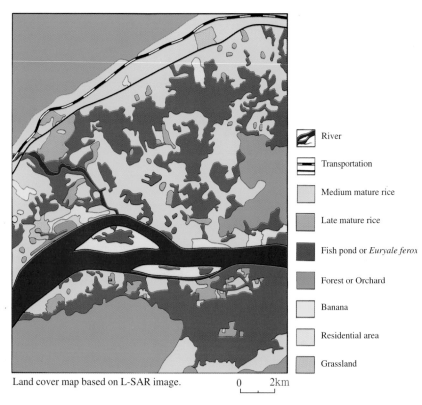

River

Transportation

Medium mature rice

Late mature rice

Fish pond or *Euryale ferox*

Forest or Orchard

Banana

Residential area

Grassland

Land cover map based on L-SAR image.

0   2km

L-SAR orthographic digital mosaic of a portion of the Sihui-Dinghu site.

Typical Land Cover Pattern in Yangtze River Delta.

The Yangtze River delta is the most developed region in China, characterized by luxuriant natural resources, highly developed industrial and transportation systems, dense population, and extensive areas of urban build-up. The alluvial plain of the Yangtze River Delta has a low elevation averaging 2 to 5m above sea level. This area is located in eastern China with a warm and wet monsoon climate and sufficient amounts of solar heat and illumination. The annual average temperature is 15 to 18°C, and annual rainfall is 1000 to 1900mm. Hazardous weather includes autumn droughts, monsoon flooding, and typhoons. Soil suitable for rice is the dominant soil in this region. Rice is the major crop, reaping two or three rice crops a year. Intensive and meticulous farming activities make it one of the highest grain production regions in China. Taixing City is at the upper center of the multi-frequency, multi-polarization false color SIR-C/X-SAR image acquired on April 18,1994. The pattern of built-up areas and the size and shape of villages clearly demonstrate three distinctive land use patterns. Bright linear features along the Yangtze River are roads or dykes with buildings or trees. Acting as corner reflectors, the buildings and trees produce strong return to radar. Villages on the east-side of the river are obviously larger than those on the west, as is field size. A regular pattern on the east-side indicates a paddy field. The small fields on the west-side are non-irrigated land. The backscatter behavior of a target is both target and radar system dependent. The five black and white single parameter radar images illustrate that different targets have a variety of signal return characteristics as a function of frequency (L, C and X band) and polarization (HH, HV and VV).

0   8km

SIR-C/X-SAR image of Taixing site (top to bottom: L-HH, L-HV, C-HH, C-HV, and X-VV).

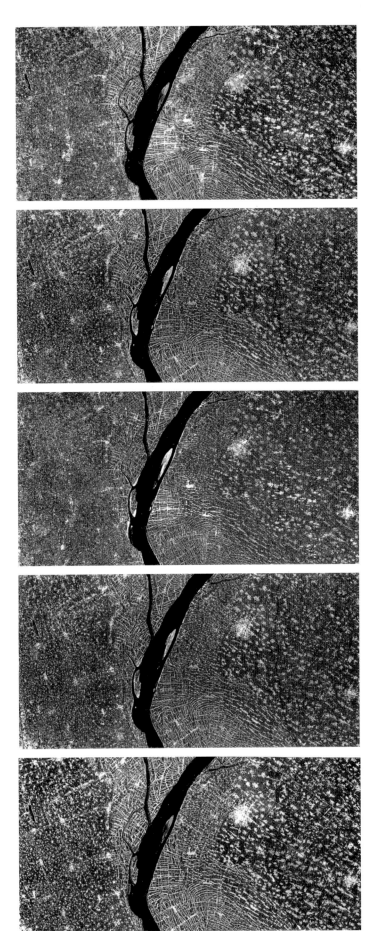

Multi-frequency, multi-polarization false color SIR-C/X-SAR image of Taixing (R: L-HH, G: L-HV, B: C-HH).

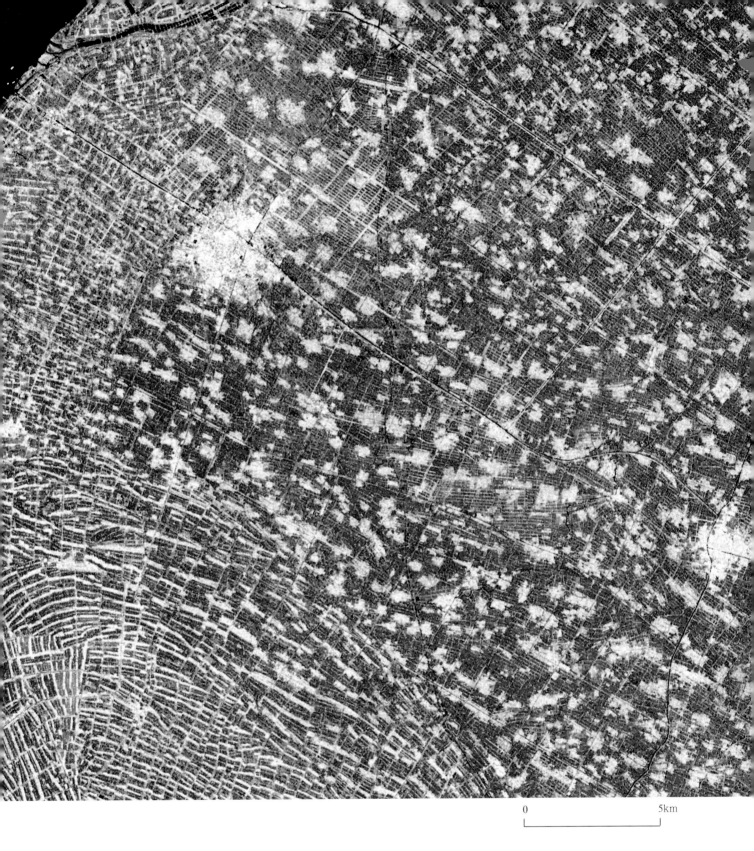

0   5km

Newly Reclaimed Land from the Sea in Lugang

A good solution for expanding land supply is to reclaim land from the sea. This is a C, L, and P band airborne polarimetric AIRSAR image of Lugang. The central latitude and longitude of this image is 24° 05' N and 120°25' E. The lozenge-shape island is newly reclaimed land from the sea. The still water, sea surfaces, and land exhibit distinctive contrast in C band image, small difference in L band, and almost no difference in P band.

As frequency becomes lower (from C band to P band), the contribution of surface cover to backscatter decreases; however, the capability of penetration increases. The comparison of backscatter behavior between an agricultural field and still water clearly illustrates this point.

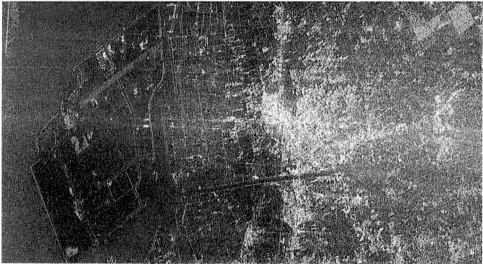

0   2km

Three band AIRSAR image of reclaimed land from the sea (top: C band, middle: L band, bottom: P band, R: HH, G: HV, B: VV).

Three band false color combination AIRSAR image of shoreline in Lugang.

0   2km

AGRICULTURE IN NORTHEAST CHINA
Beijing-Tianjin Region

The Beijing-Tianjin region is located in the Huang-Huai-Hai river plain and has a temperate semi-humid climate. Rainfall and high temperatures come in the same season. The average annual temperature is 9 to 14°C, and the annual accumulate temperature is 3800 to 4500°C for ≥ 10°C. Annual rainfall is 450 to 750mm, with 70% of the rainfall in the summer. Beijing-Tianjin has a well-developed industry, extensive transportation systems, a dense population, and areas of intensive build-up, coupled with a long history of farming activities. There are paddy fields, non-irrigated land,

land in vegetables, and irrigated land. Most of the cultivated land relies on irrigation.

This multi-frequency, multi-polarization false color SIR-C/X-SAR image was acquired on April 10, 1994. The central latitude and longitude of the image is 40°00' N and 117°08' E. The Chaobai River runs through the left corner of the image. The Yuqiao Reservoir is at the upper right corner. In early April, winter wheat has just started to develop seedlings. Various green colors in the image represent the roughness and moisture of the soil. This is typical of the high yield, irrigation agriculture of north China. Wheat and corn are major crops, reaping two crops a year.

Multi-temporal Radarsat imagery acquired in September 1997 and January 1998 is centered at 39°45'N and 116°23'E. Daxing County is on the left, and Nanyuan County is on the right. Intensively built-up areas are shown as bright patches, and agricultural fields are in dark blue. The images were collected in late autumn

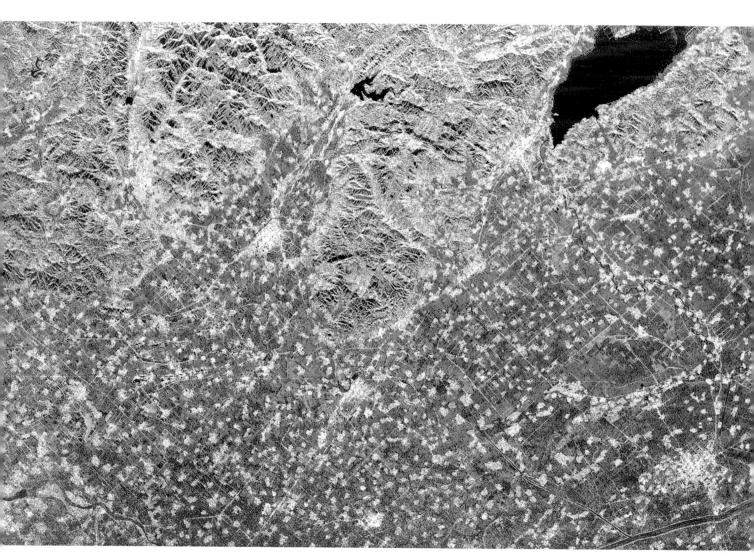

Multi-frequency, multi-polarization false color SIR-C/X-SAR image of Jixian site (R: L-HV, G: C-HH, B: L-HH).

0   10km

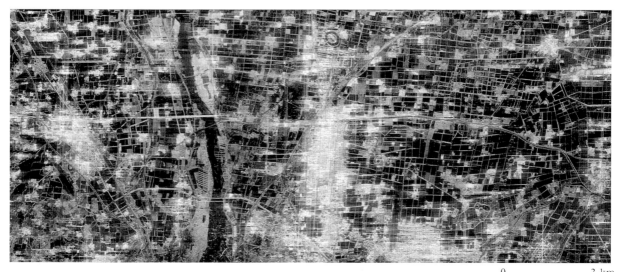

L-SAR image of Daxing site in Beijing.

0   3 km

and winter when most of the fields were bare soil, only rarely in crop. The colors shown in the image are related to the roughness and moisture of soil. Only limited information is related to crops.

The L band, HH polarization airborne L-SAR image of Daxing area was acquired in October of 1997. This image, with high resolution, clearly delineates the boundary of agricultural fields with shelter forest belts. Cornfield is in gray tone and the bare soil is in dark tone.

Multi-temporal Radarsat image of southern Beijing.

0   3 km

Bohai Coast

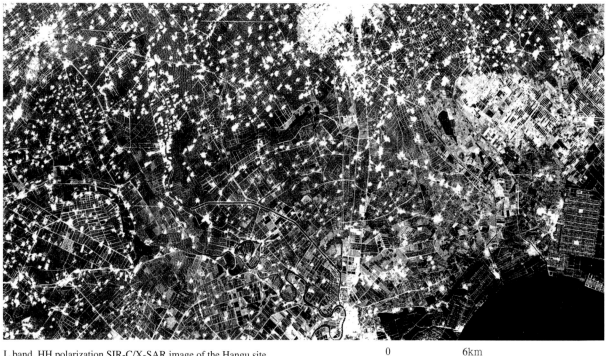

L band, HH polarization SIR-C/X-SAR image of the Hangu site.

0   6km

The L band, HH polarization SIR-C/X-SAR image of the Hangu site illustrates, from west to east, the landform changing from alluvial plain to shoreline, land use changing from agriculture to aquaculture, from crop farming to fish and shrimp farming and salt production. Hangu City is at the upper center of the image. On the west of Hangu City, land is mainly used for agriculture; on the east of Hangu City, land is used for aquaculture and salt evaporation. The regular, grid pattern fields at the right corner are salt pans, fish ponds or shrimp ponds. This is typical land use pattern of agriculture and aquaculture in northeastern China.

AGRICULTURE AND VEGETATION IN NORTHWEST SEMI-ARID REGION

Erqisi River Basin

The Erqisi River catchment lies in between the Altay Mountains on its north and the Jungarr Basin on its south and has a continental temperate arid climate. The average annual temperature is 6 to 10°C, the annual accumulate temperature is 3100 to 3600°C for ⩾ 10°C, the annual rainfall is about 200mm, and the annual sun illumination

is 2600 hours. Winters are long and cold. The large amount of snow cover helps crops to survive over the winter. Summers are hot. The large temperature difference between day and night is favorable for crops to accumulate nutrients. The region has a good supply of surface and underground water and fertile soils. There are sufficient land resources for irrigated agriculture and livestock farming. Spring wheat, oil crops, and forage grass predominate.

A multi-frequency, multi-polarization false color SIR-C/X-SAR image of the Buerjin River within the Erqisi River catchment was acquired on April 11, 1994 and centered at 87°12'E, 47°46'N. The Buerjin River shows up as a bright meandering pattern running through the image from east to west. The trees along the river generate strong backscatter. Three canals run from the Buerjing River to the right corner of the image. Forest belts in a regular grid pattern are discernible on both sides of the river. The forest belts represent the distribution of canals. The cultivated lands lay within the networks of the irrigation canals. Within the Erqisi catchment, crops are totally dependent on irrigation. These cultivated lands have been reclaimed from the Gobi Desert and survive in very adverse conditions. Snow melt from the Altay Mountains provides sufficient water to cultivate the lands in the summer, a factor that has promoted the development of agriculture in this region in the last five decades.

A multi-frequency, multi-polarization false color SIR-C/X-SAR image of the Kelan River within the Erqisi River catchment area was acquired on April 11, 1994 and centered at 87°49'E, 47°33'N. The Kelan River, which originates in the Altay Mountains, runs through the image from top to bottom. Vegetation is well developed along the river. The backscatter from vegetation is much stronger than the surrounding Gobi Desert. Forest belts along the irrigation canals in grid pattern lay on both sides of the river. The Kelan River, carrying snowmelt from the Altay Mountains, runs into the Jungarr Basin. Within its catchment, local residents draw water from the river for irrigation. Irrigated agriculture on the alluvial fans is well developed in this region. The irrigation canals show up as bright linear features due to trees lining the canals generating strong backscatter. The surrounding desert areas in dark green are anticipated to be areas suitable for future development.

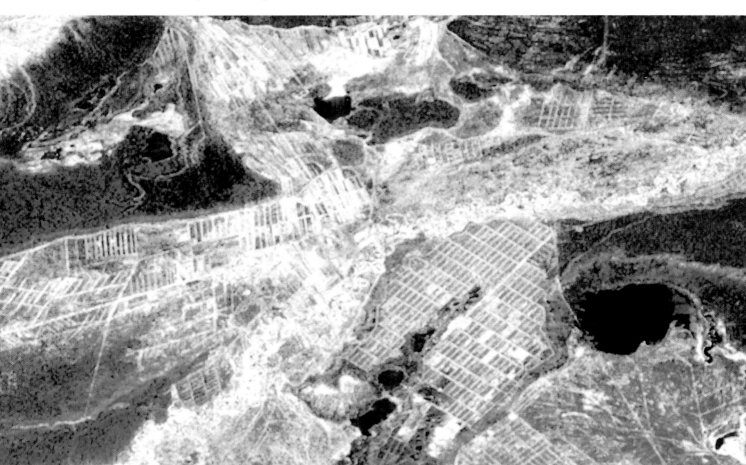

Multi-frequency, multi-polarized false color SIR-C/X-SAR image of the alluvial plain in front of the Altay Mountains (R: L-HV, G: C-HV, B: L-HH).

0   5 km

0   5km

Multi-frequency, multi-polarized false color SIR-C/X-SAR image along the Kelan river in front of the Altay Mountains (R: L-HV, G: C-HV, B: L-HH).

Tarim River Catchment

The X band, VV polarization SIR-C/X-SAR image of the Tarim River catchment was acquired on April 10, 1994. Two segments of image have been selected, one is centered at 41° 15'N and 79° 44'E. The site is located west of Tacheng City of Xinjiang, with the Tianshan Mountains on its north and the Tarim Basin on its south. The Tarim River catchment has a temperate climate, with an average annual temperature of 9.8℃. The annual accumulate temperature is 3910℃ for ≥ 10℃, and the annual rainfall is only 58.4mm. This region is suited for heat tolerant crops like rice and cotton, and produces one crop a year. The Tarim River carries snowmelt derived from the Tianshan Mountains and provides sufficient water for local residents, their crops and animals. Population in irrigated land areas is denser than other areas. The Aksu Plain is an important grain and cotton production base in Xinjiang. It is interpreted from the radar image that regular fields are cultivated lands, and irregular fields are reclaimed pasture. The image was acquired in early spring when the fields were being irrigated in preparation for cultivation. The tonal variation of the image indicates the moisture variation of each field.

The lower image is centered at 40°34'N and 80° 37'E, on the margin of the Tarim Basin. The Takla Makan Desert is on the right side in dark tone. The Aksu River and Shangyou Reservoir provide enough water supply for farming activities. The Shangyou Reservoir is at the center of the image. The bright features along the reservoir are trees, grasses and reeds. The regular patches at the upper center of the image are cultivated lands to grow rice and cotton. The image was acquired in early spring. At that time, fields were being irrigated to prepare for cultivation. Paddy fields are ready for transplanting. The irrigated fields show up as bright patches because they have a high water content. The moisture generates a high dielectric constant, which results in strong backscatter coefficients. The dry fields may have low backscatter coefficients. On the left corner of the image, there are pasture lands. The local residents breed and raise many animals on the large, wide-open, beautiful pastures.

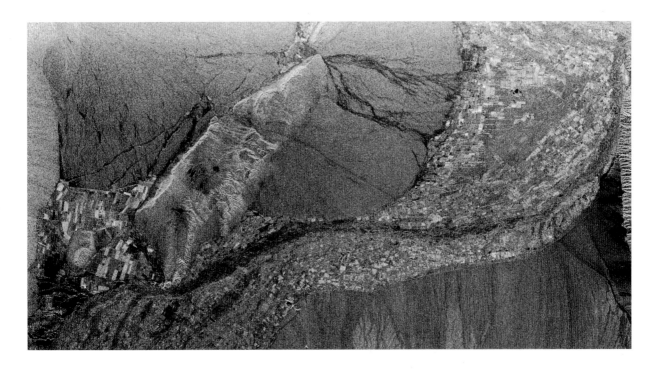

0   15km

X band, VV polarization image of the Aksu Plain in the Tarim Basin.

Desert Vegetation

The image of SIR-C in the region of desert vegetation was imaged approximately from northwest to southeast, that is from Tao Taole County which is located in the center of the Yin Chuan Plain to Dingbian County near the Loess Plateau. The image covers an area about 200 km in length and 50 km in width. The center of the image is 37° 55.6'N and 107°55.6'E. Some counties included in the image are: Dingbian, Yanchi and Otog Qiangi.

The false color SAR image was composited using L-HV, C-HH and C-HV, three polarizations of C and L bands, and are assigned red, green and blue respectively. The SAR systems are sensitive to soil moisture and differences in object structures on the ground. Some land cover types, such as bare sand, town, and salt lake, are easily identified. Some vegetation growing on plantation and dune , such as *Artemisia ordosica*, *Nitraria sibirica*, and *Achnatherum splendens*, are identified directly on SAR images. On this SAR image, sand dunes are dark green with purple speckles and are identified easily by their rough structure or texture. *Artemisia ordosica* community associated with the sand

False color multi frequency, multipolarized composite image of SIR-C/X-SAR in the desert vegetation region of Taole to Dingbian.

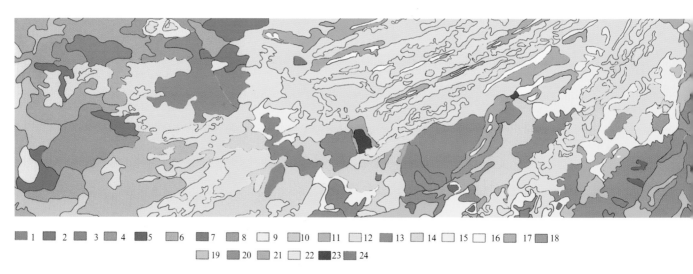

■ 1 ■ 2 ■ 3 ■ 4 ■ 5 ■ 6 ■ 7 ■ 8 □ 9 ■ 10 ■ 11 ■ 12 ■ 13 ■ 14 □ 15 □ 16 ■ 17 ■ 18
■ 19 ■ 20 ■ 21 □ 22 ■ 23 ■ 24

Classification map for vegetation in the desert vegetation region.

dunes is light blue with a few purple grains, and it is also identified easily. The majority of lowlands between the dunes are Solonchak and Solonetz soils where the groundwater is near the surface of land, and in the image the majority of these areas are various blue colors, in which dark blue represents *Nitraria sibirica* communities and light blue or light green represent *Achnatherum splendens* communities. Farmland can be identified according to the texture of the grid form. Other vegetation types are identified according to the TM images and a variety of maps and documents.

On three Landsat-5 TM false color images that were acquired from July to September in 1987, sand is light brown, Artemisia ordosica communities growing around sand is dark brown, the Solonchak and Solonetz soil between dunes is white and dark gray, and the vegetation growing on this soil represents mainly *Nitraria sibirica* communities, with infrequent *Achnatherum splendens* communities which are red with a striped shape. Farmland is also identified according to the texture of the grid form.

In the image of NOAA Satellite, vegetation types, which are identified in the images of SAR and TM, can not be identified except for the dunes, which have high reflectance and are distributed, with a long striped patten due to the lower resolution of NOAA.

0   15km

1. *Stipa bungeana steppe*
2. *Stipa grandis steppe*
3. *Stipa breviflora steppe (includes Agropyron mongolicum, Stipa bungeana, etc.)*
4. *Thymus serphyllum steppe*
5. *Artemisia frigida steppe*
6. *Stipa breviflora steppe (includes Reaumuria soogorica, Salsola passerina, Ajania achilloides, Iris bungei, etc.)*
7. *Ajania achilloides steppe*
8. *Oxytropis aciphylla xerophily weed steppe*
9. *Reaumuria soogorica, Salsola passerina desert*
10. *Salsola arbuscula desert*
11. *Pegganum nigellastrum desert*
12. *Artemisia ordosica community (includes Glycyrrhiza uralensis, Pennisetumcentrasiaticum,*

Sophora alopecuroides, etc.)
13. *Glycyrrhiza uralensis community (includes Agropyron mongolicum, Sophora alopecuroides, etc.)*
14. *Sophora alopecuroides community (includes Glycyrrhiza uralensis, Pennisetum centrasiaticum, etc.)*
15. *Pennisetum centrasiaticum community*
16. *Nitraria sibirica scrub (includes Achnatherum splendens)*
17. *Achnatherum splendens community*
18. *Aneurolepidium dasystachys community*
19. Spring wheat
20. Broom corn and millet
21. *Caragana korshinskii*
22. Sand
23. Water
24. Town

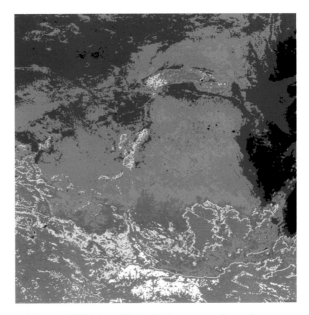

The image of NOAA satellite in the desert vegetation region.

There are three vegetation zones, which are Typical Steppe, Desert Steppe and Steppe Desert and which progress in order from right (southeast) to left (northwest) on the map produced depending on the interpretative results of the images of SAR, TM and NOAA. There are various and complex vegetation types because three vegetation zones exist and a lot of dunes are distributed in the region of the map.

The Landsat TM image in the desert vegetation region.

AQUACULTURE

The Bohai Bay

This X band, VV polarization SIR-C/X-SAR image of the Tangshan site illustrates typical aquaculture land use in the north of China. On the left part of the image is a typical agricultural land use pattern of north China. On the right part of the image is a typical aquaculture season. This region is favorable for aquaculture such as fish, shrimp, and shellfish farming. It produces high quality prawn, shellfish and laver. On the right part of the image, water ponds, in regular grid pattern, are aquaculture farms. On the left part of the image, forest belts show up as thin bright linear features. Paddy rice fields lay within the forest belts. Trees and buildings are constructed along the roads or channels in this region. Because of the heavy rainfall and flat landform, there is a need for efficient drainage for this region. This is a typical tideland aquaculture land use pattern in southern China.

X band, VV polarization image of the Tangshan site.

0   10km

land use pattern of north China. The dominant soil type in this region is a saline-alkali soil. The vegetation is almost entirely salt-resistant natural vegetation.

Yangtze River Estuary

The X band, VV polarization SIR-C/X-SAR image of the Yangtze River Estuary was acquired on April 18, 1994 and is centered at 32°53'N and 120°47'E. This region is very flat with an average elevation of 2 to 5m. This region is characterized by a subtropical humid monsoon climate. It has luxuriant water resources and a dense drainage system organized in a regular grid pattern. The average annual temperature is 14 to 15°C, annual accumulate temperature is 4450 ~ 4770°C for ⩾ 10°C, and the annual rainfall is 1000 to 1100mm with concentrations in the summer

X band, VV polarization image of the Yangtze River Estuary.   0   10km

FORESTRY

Remote sensing studies of forests are important for understanding phenomena such as the global carbon cycle, hydrologic cycle, and energy balance. In 1983 NASA estimated that forests cover 33% of the total land surface, contain 90% of the standing biomass, and yield 65% of the net primary production. An increase in both global population growth and deforestation around the world requires accurate and timely information about the distribution and rate of change of global vegetation. This information has traditionally been derived from a variety of sources, the accuracy of which is difficult to ascertain. Remote sensing is a potential method for gathering this information, as has been demonstrated from AVHRR data. However, pervasive fog and cloud as well as limitations in solar illumination in the tropics and subtropics and high latitudes severely limit the utility of visible and infrared sensors. Microwave remote sensing, with its near all weather, day and night capabilities, is a powerful tool for acquiring biophysical data that will improve understanding of global climate, the hydrologic cycle, the carbon-nitrogen cycle, and the global energy balance.

This part presents the results of studies of forest discrimination, classification, and volume estimation in the northern, southern, and southwestern portions of China using airborne and spaceborne imaging radar data. It further demonstrates that SAR data have forestry applications and the potential for contributing to ecological research.

FOREST DISCRIMINATION AND CLASSIFICATION

Forest Land Discrimination
Altay Area, Xinjiang

The Altay test site, an area of mountains and plains, is located in the most northern portion of Xinjiang Autonomous Region. The mountainous areas have abundant rainfall, high winds, and a short frost season influencing the distributions of dense canopy and grasslands with obvious vertical zonality. The plains areas exhibit the characteristic distribution of desert steppe, steppe, and desert from the base of the mountain to the basin center due to less rainfall. Only the edges of gullies and alluvium fans, and selected lower topographic regions have ecological conditions capable of supporting trees, shrubs, and agriculture.

The Shuttle Imaging Radar (SIR) data operating in Modes 16 and 11 were collected at the site in October of 1994. The Mode 16 data are useful for discriminating vegetation in the river valley, mountainous steppe, and forest belt in the plain along the Burqin River and its

surrounding regions. The Mode 11 data are best for determining vegetation in the river valley along the Ertix River.

The multifrequency and multipolarization images composed by Mode 16 data show that deciduous forests dominated by poplar have a different color from dry prairie. These deciduous forests are found in the Burqin River valley and its northern regions. The regular grid pattern of forest belts around the farmlands is clearly shown in the images as a white color. In the region east of the river valley, reed lands are blue and blue-green. The very bright, almost white, triangular shaped ground in the upper right image is bedrock. In the six single band and polarization images, deciduous vegetation in river valleys and forest belts can be easily discriminated in L and C bands and all polarizations images; however, reed land is not clearly displayed in L band or any single polarization image, especially in the L-HH image. Bedrock and dry mountain steppe also can not be discriminated in the single band and single polarization images.

The false color image composed by Mode 11, L band and C band, and HH and HV polarizations shows deciduous forests distributed in the Extix River valley as bright white, and marshlands as a green zone nearby.

R: L-HH, G : L-HV, B : C-HV

0   8km

False Color composite images of multifrequency and multipolarzation SIR-C data in the Altay area, Xinjiang (Mode 16)

R: L-HH, G : L-HV, B : C-HH

False Color composite images of multifrequency and multipolarization SIR-C data in the Altay area, Xinjiang (Mode 11, Red: L-HH, Green: L-HV, Blue: C-HV).

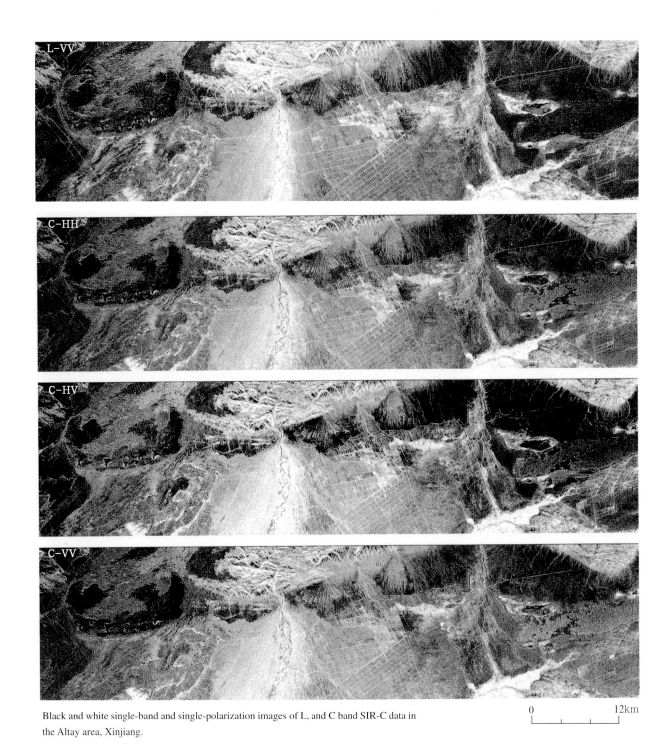

Black and white single-band and single-polarization images of L, and C band SIR-C data in the Altay area, Xinjiang.

0   12km

Tarim River Area

This test site is located in the north of the Tarim Basin, and has a low and flat terrain and well developed river meanders. The average annual temperature is 10°C. The frost-free season length is 180 to 210 days. Annual rainfall is 25 to 35mm. The Tarim River brings fertile soils to the area so that a 500km long vegetation zone called the "green corridor" has been formed along the river, which consists of *Populus diversifolia*, willow, bush and reed , etc.

Mode 11 SIR-C/X-SAR data acquired in October of 1994 were used to study the area and to produce the false color images. The false color images, composed by L-HH (R), L-HV (G) and C-HV (B), show that *Populus diversifolia* forests distributed in the Tarim River valley appear bright white in color. This means they have strong radar backscatter in all three band and polarization combinations used to generate the false color composite. This is probably due to the 10m height of poplars and willows with more backscatter from underbrush. The marshlands exhibit a green color outside the zone of *Populus diversifolia*. Deserts are dark red and pink color due to weak radar backscatter, especially in L-HV and C-HV.

False color composite SIR-C image shows distribution of *Populus diversifolia* forests in the Tarim River Valley.

False color composite SIR-C image shows distribution of marshlands in the Tarim River area.

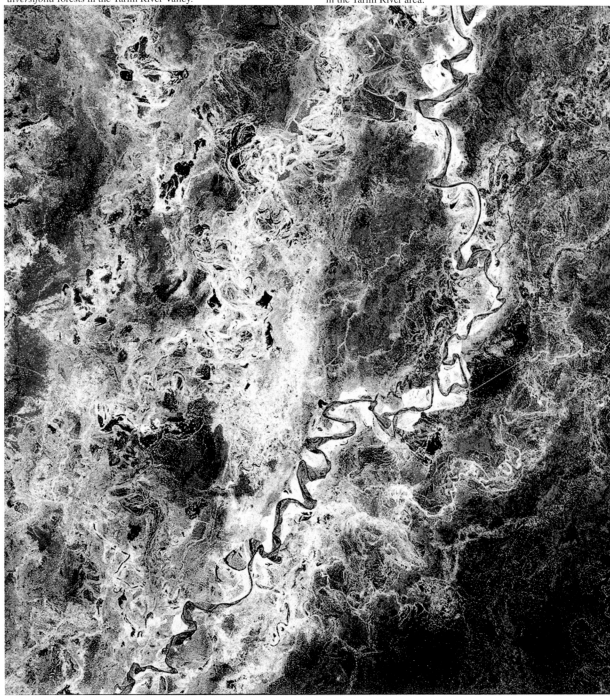

False color composite SIR-C image of the Tarim River area (Red: L-HH, Green: L-HV, Blue: C-HV).

0   8km

Northern Area of Yinchuan

The test site is located east of the Helan Mountains, west of the Yellow River, and north of Yinchuan City. It has a variety of geomorphologic types, including desert, sand dunes,, mountains, plateaus, river plains, valleys, and basins. The river plains, valleys, and basins are characterized by flat terrain, a thick soil layer, and intensively cultivated lands. Mountainous regions have high elevation and an obvious vertical zonality of vegetation. Desert, sand dune, lake and basin, and grass are crosswise distributed. The test site is in a dry to semi-dry climate zone. Rainfall is not seasonally well distributed, and there is high wind duration and velocity as well as blowing sand. Mixed forest and bush are found in the mountainous regions. The river plains are important areas for irrigated agriculture.

Shuttle imaging radar data operated in Mode 11, L and C bands and HH and VV polarizations were acquired in April of 1994 and were used to study this site. The data were processed and composited into false color multifrequency and multipolarization images.

In the false color image composited from L-HH (R), L-HV (G) and C-HV (B) channels, mixed forests and bush are a bright green color and are distributed in the Helan Mountain region located in the west portion of the false color radar image. Farmlands located in the eastern portion of the image are mainly flat paddy fields. The thin forest shelterbelts are crosswise distributed between paddy fields and show a bright yellow color with red speckle in the image. Where deserts have encroached on some farmlands due to the lack of forest belts, they appear as a dark red color on the image.

False color composite imagery of SIR-C/X-SAR data in the northern area of Yinchuan City (Red: L-HH, Green: L-HV, Blue: C-HV).

0   8km

False color composite imagery of SIR-C/X-SAR data shows forest belts around farmlands.

False color composite imagery of SIR-C/X-SAR data shows deserts encroaching on the farmlands.

Pinggu Area in Beijing

The Pinggu area is located in the northeast of the city of Beijing with an elevation of 300 to 800m in its north, and an elevation of 10 to 50m in its south. The site has a temperate semi-humid climate; maximum rainfall and most of the heat supply come simultaneously in the summer. Forests mainly consist of coniferous and shelter forests. The coniferous forests are dominated by pines distributed in the lower mountains and hilly regions, and the shelter forests are dominated by white poplar distributed around the farmlands. Grasslands are scattered in the lower slopes.

Shuttle Imaging Radar (SIR) data acquired in Mode 11, L and C bands and HH and VV polarizations were collected during October 1994. The color image composed by L-HH (R), L-HV (G) and C-HV (B) channels exhibit the coniferous forests in the mountainous region as a bright green color. Hills and grasslands located between the mountains and plains are brownish red. The thin, linear forest shelterbelts around farmlands have a strong radar response and appear as a bright yellow and white color in the image. Reed lands are denoted as patch and linear shapes on both sides of the river and are bright red in the image.

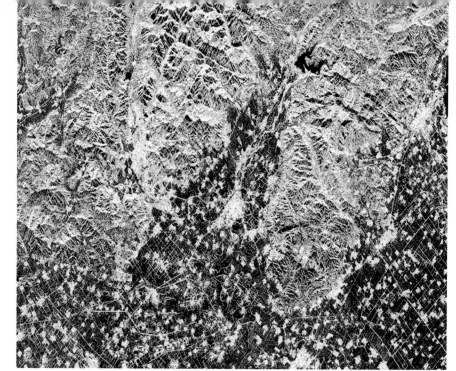

False color image composed by SIR-C/X-SAR data in the Pinggu area (Red: L-HH, Green: L-HV, Blue: C-HV).

0   8km

Multifrequency and multipolarization image shows reed lands (bright red) along two sides of the river in the Pinggu area.

Multifrequency and multipolarization image shows coniferous forests (bright green) and grasslands (brownish red) distributed in hilly and level slopes in the Pinggu area.

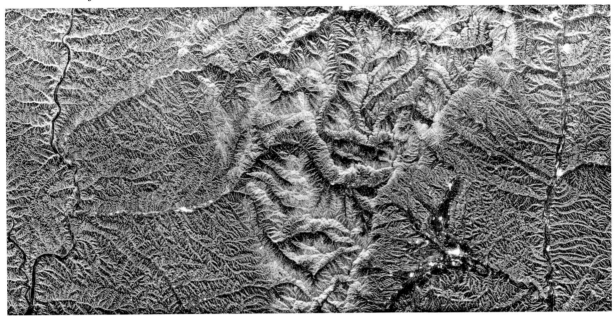

SIR-A black and white image of the Luya mountain area in the northern part of Shanxi Province.

The test site is located in the Luya Mountain area in the northern portion of Shanxi Province. The average elevation is 1800m. The landforms vary from high relief, coarse textured landscapes in the center to lower but highly dissected loessial areas on the left and right sides of the image. It has a continental monsoon climate with high winds and large amounts of sand. The Fenhe and Huanghe Rivers pass though the area in the east (right) and west (left), respectively, and form temperate forest steppe regions on the Loess Plateau. The forest type is mainly deciduous such as Liaodong oak, white birch and poplar.

The SIR-A image of this site shows the Luya Mountain in the middle of the image with an obvious black and white boundary. The mountains consist of metamorphic rock. Dense deciduous stands growing to an elevation of 1500m are white due to strong radar response, and the river valley is dark grey. Soil has little erosion because vegetation cover inhibits surface runoff.

Forest Type Discrimination And Classification

Three North Shelter Forests in the Yichuan Area of Shaanxi

A test site for three north shelter forests was selected west of Yichuan County, Shaanxi Province, with the Yellow River flowing south on the eastern side of the image. This site has an average elevation of 1300 to 1400m, with hilly landforms of loess characterized by gullies and ravines due to surface erosion. It is a temperate, dry to semi-dry region with little annual rainfall.

The forest in the site is part of three north shelter forests in China. Forest types are mainly deciduous, coniferous, and mixed forests. Deciduous trees are mainly poplar and oak, coniferous trees are pine stands, and mixed forests are composed of deciduous mixed forest and coniferous-deciduous forest. In April of 1994, SIR-C/X-SAR operating in Mode 11 imaged the test site, and false color frequency-polarization combinations of L-HH, L-HV, C-HH, C-HV, and X-VV were generated. We performed geometric rectification and speckle filtering prior to generating the false color composite images.

In the false color composite image of L-HH (Red), L-HV (Green) and C-HV (Blue), coniferous forest appears as yellow, mixed forest as red and white, and deciduous forest as blue; therefore three forest types can easily be discriminated by using color. However, in the single band and single polarization images, it is very difficult to identify and separate these three types of forest.

A supervised classification, maximum likelihood method was employed using L and C bands, HH and HV polarizations. Coniferous, deciduous, and mixed forest, river course plain or floodplain, and radar shadow can be discriminated on the image. The classification results showed that coniferous forest had the best classification accuracy at 79.7%, mixed forest had an accuracy of 68%, and deciduous forest had an accuracy of 60.2%. The average classification accuracy was 75.10%, and the overall accuracy was 70.4%. Therefore, SIR-C/X-SAR data were quite successful for the discrimination and classification of three north shelter forest types.

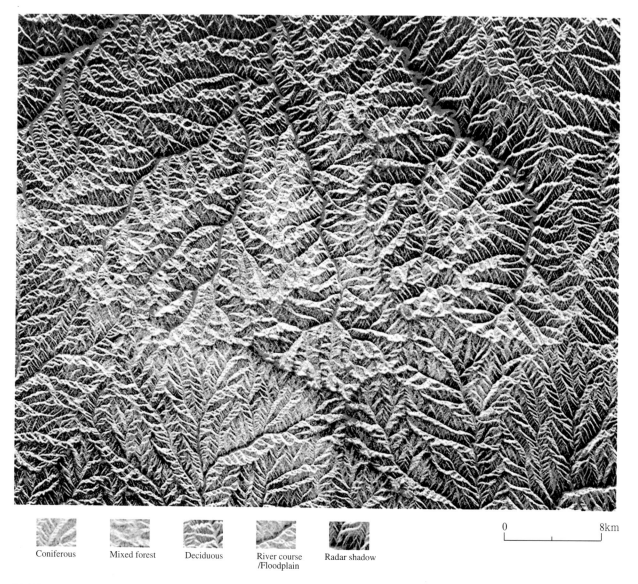

Coniferous	Mixed forest	Deciduous	River course /Floodplain	Radar shadow

0   8km

Multifrequency and multipolarization false color SIR-C/X-SAR image of three north shelter forests in the Yichuan area, Shaanxi Province.

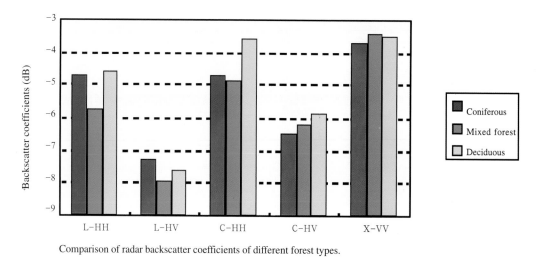

Comparison of radar backscatter coefficients of different forest types.

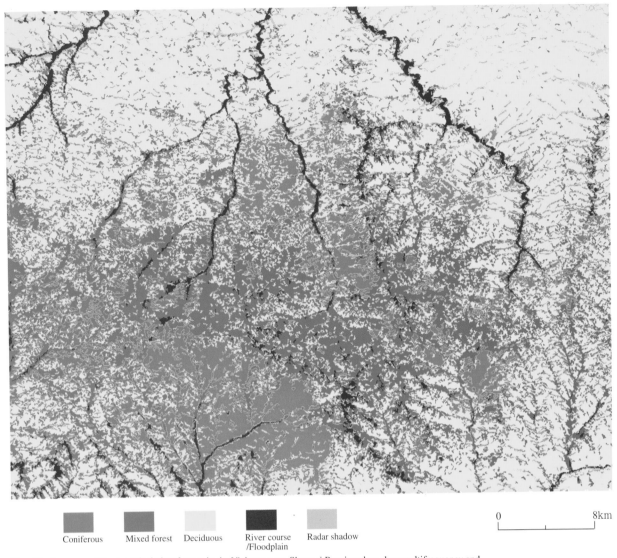

Coniferous   Mixed forest   Deciduous   River course /Floodplain   Radar shadow

0 ——————— 8km

Classification map of three north shelter forests in the Yichuan area, Shaanxi Province based on multifrequency and multipolarization SIR-C/X-SAR data.

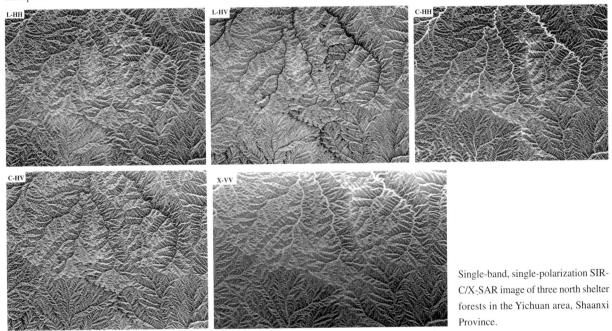

Single-band, single-polarization SIR-C/X-SAR image of three north shelter forests in the Yichuan area, Shaanxi Province.

47

Coastal Shelter Forests and Tropical Crops of Hainan

This test site is located in the Leizhou Peninsula and Hainan Island of the South China Sea. Topographically, Leizhou Peninsula is mainly a terrace with the highest elevation less than 100m. Hainan Island is characterized by asymmetric ring and layer structures, changing from mountain and hilly to terrace, and eventually to coastal plain. This region has a subtropical and tropical monsoon climate with high sun illumination. Sufficient heat and water supply is favorable for vegetation growth. The vegetation is comprised of plantation shelter forest, fuel forest, bush, tropical crops, and orchard.

Multifrequency and multipolarization SIR-C/X-SAR data were acquired in April of 1994. The false color images of Leizhou Peninsula clearly show coastal shelter forest as green and greenish blue and different tropical crops as yellow and light yellow with red patches. A large blue area in the eastern portion of the image is flat, non-irrigated land.

The false color images of Hainan Island show that Haikou City has the strongest radar return (brightest). Coastal shelter forests, located in the north of this city, are green and green with blue. In the image composed by L-HH (R), L-HV (G) and C-HH (B) channels, the large green area in the western portion of the image is fuel forest and bush, and the boundary between shelter forest and farmland is very clear. However in the image generated from L-HH (R), L-HV (G) and C-HV (B) channels, fuel forest and bush appear as light blue, and the boundary between shelter forest and farmland is not as clear. Different types of tropical crops appear as yellow, bright yellow with white, and red patches. In the single band and single polarization images, fuel forest can be clearly identified on L-HV imagery. The boundary between farmland and coastal shelter forest is very clear on the C-HH image, but can not be identified on the C-HV image.

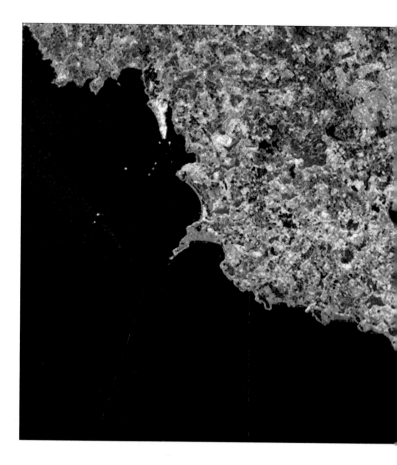

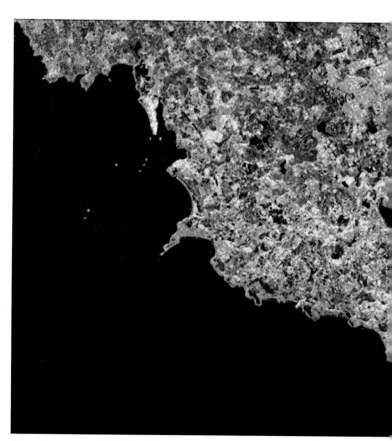

False color composite SIR-C/X-SAR images of the Leizhou Peninsula.

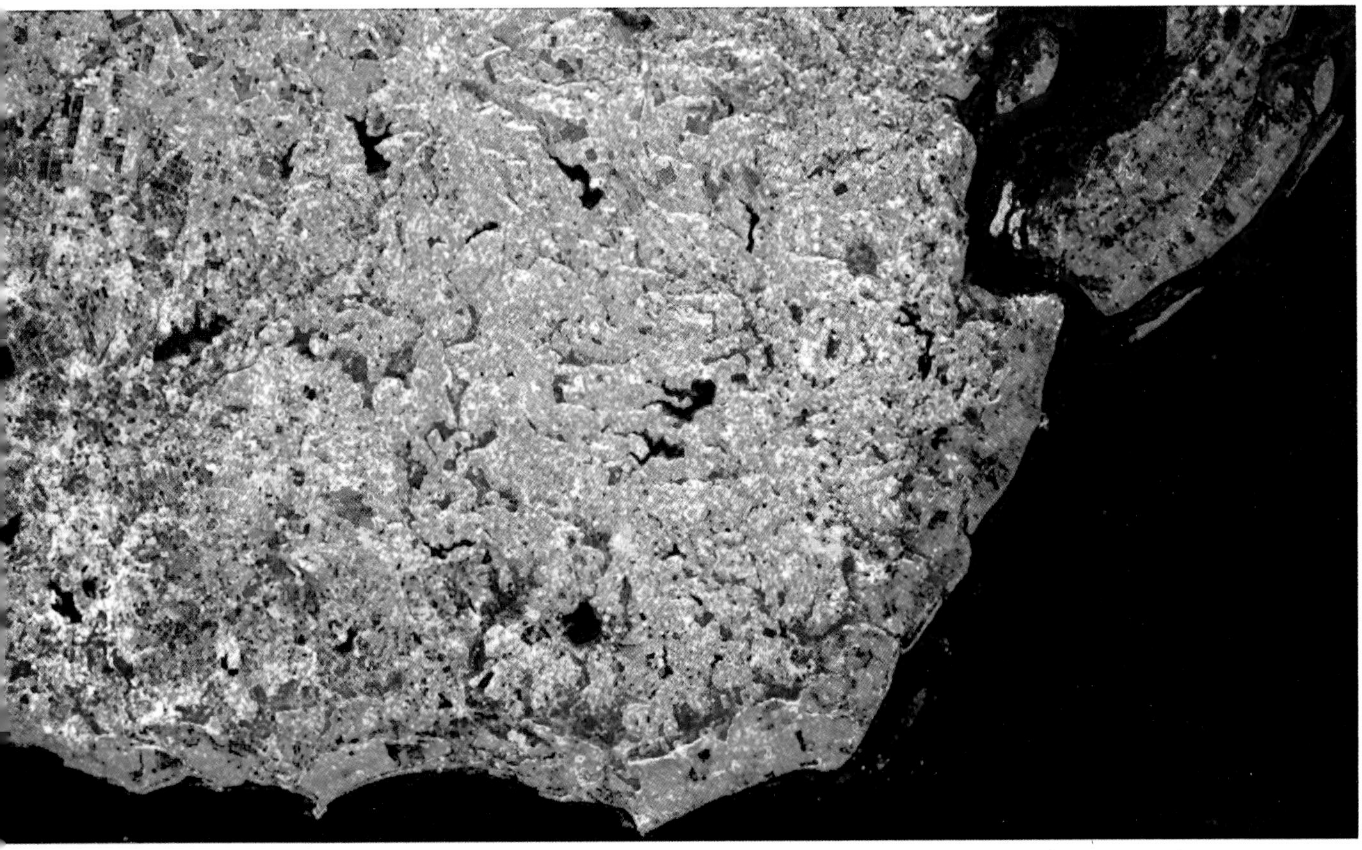

R: L-HH, G : L-HV, B : C-HV

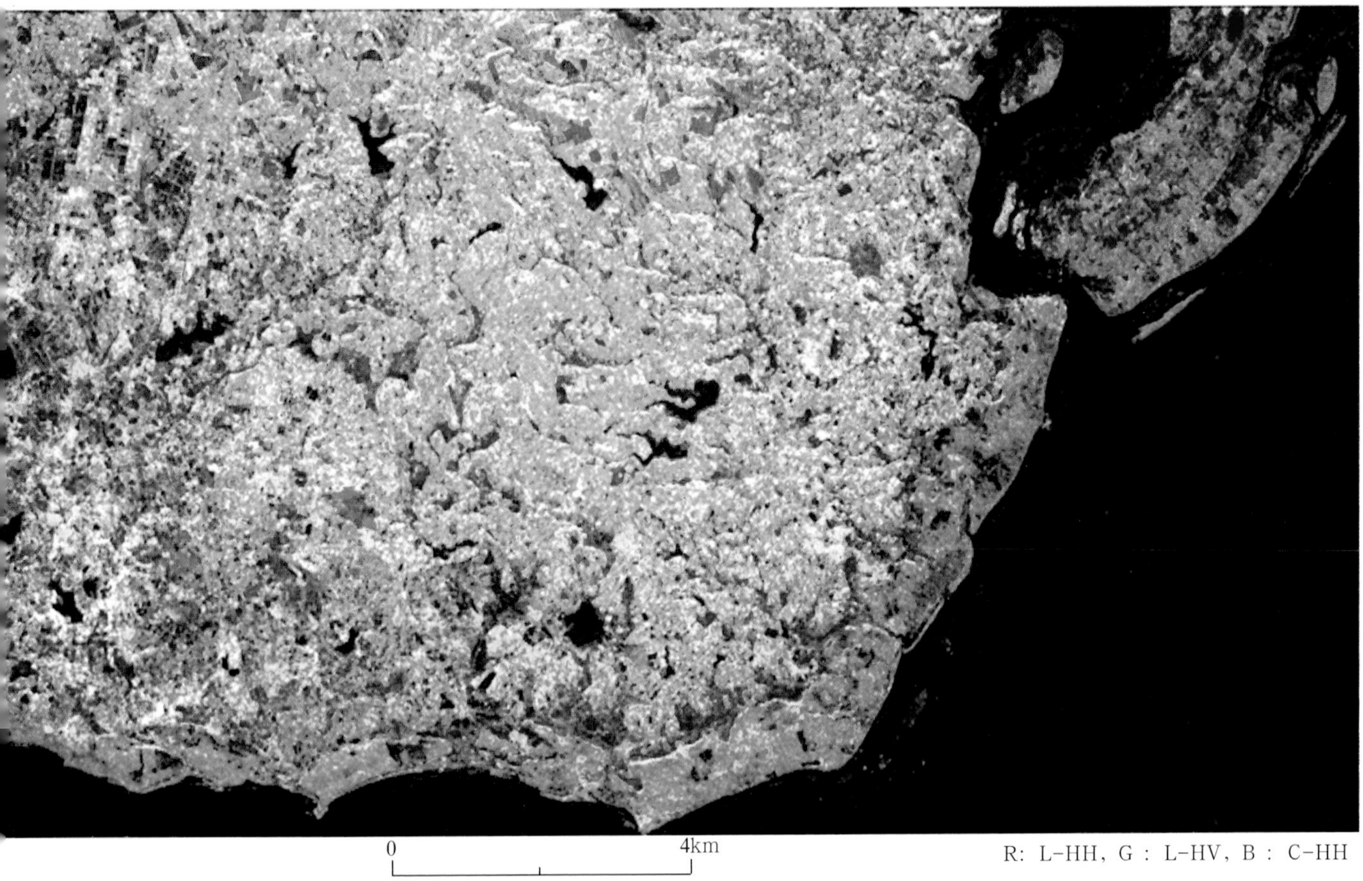

R: L-HH, G : L-HV, B : C-HH

False color composite multifrequency and multipolarization radar images of the northern region of Hainan Island
(R: L-HH, G: L-HV, B: C-HH).

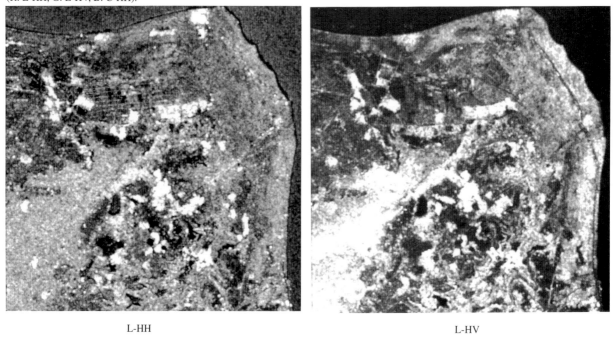

L-HH

L-HV

L band SIR-C images of the northern region of Hainan Island.

False color composite multifrequency and multipolarization radar images of the northern region of Hainan Island (R: L-HH, G: L-HV, B: C-HV).

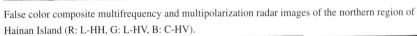

0   4km

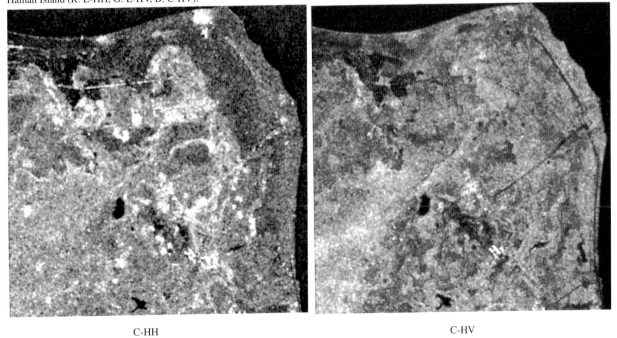

C-HH   C-HV

C band SIR-C images of the northern region of Hainan Island.

Coniferous and Deciduous Forests in Zhaoqing

The Zhaoqing test site is located in a subtropical and tropical region of southern China. The C and X band multipolarization SAR data acquired during the GlobeSAR mission in November of 1993 are useful for forest discrimination.

We performed antenna pattern corrections for the data to remove the artificial brightening in range due to side lobeing illumination. The digital elevation model (DEM) generated by digitizing a 1:50,000 topographical map was used for geometric correction of the data and to reduce the terrain effect. Combining C and X band with HH, HV and VV polarization SAR data produced the false color image. Filter and image enhancement techniques were used to improve the quality of image.

The SAR image could be shown in the 3-D viewer after it was geocoded to the DEM.

The false color image, generated by C-HH (R), C-HV (G) and X-VV (B) channels, shows that forests have a different tone and texture from other targets (such as water body, farm land, or orchard). Pine and eucalyptus stands can be discriminated based primarily on their different tones in the image. The eucalyptus stands with wide and broad leaves produce stronger radar backscatter than the pine with needle-type leaves. This further demonstrates that the canopy contributed to the radar backscatter in C and X bands, and that HV polarization is the most favorable polarization scheme for discriminating forest types.

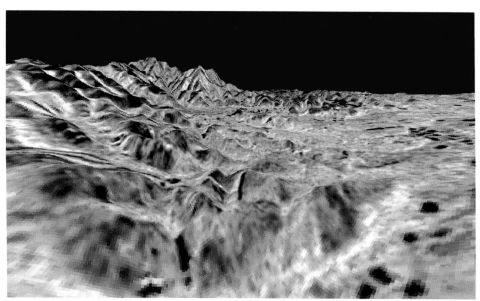

The 3-D GlobeSAR image of the Zhaoqing area.

 Eucalyptus

Pine

 Orchard

Farm land

 Water body

False color composite GlobeSAR image in the north of the Sihui area of the Zhaoqing test site (R: C-HH, G: C-HV, B: X-VV).

FOREST VOLUME ESTIMATION

Pine Forest Volume in Zhaoqing

We extracted the radar backscatter intensity from GlobeSAR data of the C and X bands for both HH and VV polarizations from the Zhaoqing test site, where ground inventories had been conducted. A multiple linear regression fit was then performed to derive an expression relating to the natural logarithm of the forest volume. The regression is of the form

$$y=a_0+a_1x_1+a_2x_2+a_3x_3+a_4x_4$$

where a_0, a_1, a_2, a_3, a_4 are the regression coefficients; y and x are forest volume and backscatter intensity, respectively; and x_1, x_2, x_3, and x_4 denote the radar backscatter intensity values at C-HH, C-VV, X-HH and X-VV channels. The regression coefficients were computed to obtain: a_0=-10.6589, a_1=-0.13, a_2=-0.66, a_3=1.20, and a_4=0.59. The regression algorithm was used to estimate the pine forest volume within the 600 hectare area of the study site.

Predicted forest volume levels from the radar data were compared to the actual volume levels. We selected three sub-areas to compute the errors between the predicted and the actual volumes expressed in percent of actual volume. For these sub-areas, the errors in the predicted volume are 15.4%, 2.9% and 10.3%. These errors are very low and are comparable to the precision of the surface estimates. These results demonstrate the potential of multifrequency and multipolarization SAR data for estimating forest volume.

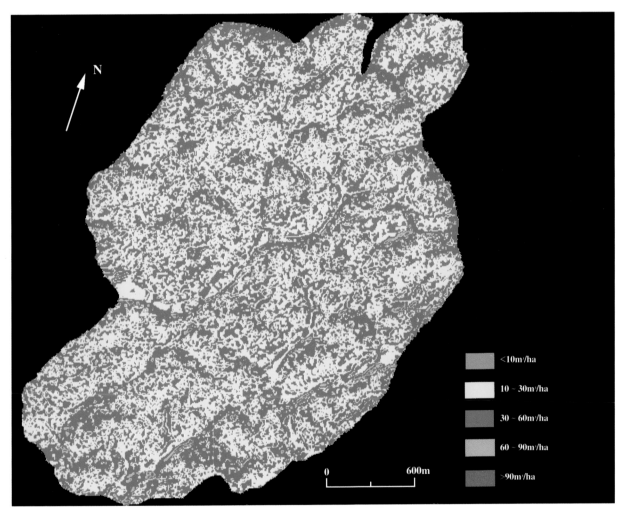

Pine forest volume map produced by the GlobeSAR data at the Zhaoqing test site, Southern China.

Forest Volume of Luoshan in Shandong

The Luoshan Forestry Centre is located in the northeastern Zhaoyuan City of Jiaodong Peninsula and has a temperate monsoonal climate. The Centre includes an area of 473.9ha where more than 80% of the tree stands are pines planted during 1956 to 1958.

Data from ERS-1 SAR, JERS-1 SAR and Landsat TM were used in the study. They were acquired in April of 1992, May of 1993 and September of 1992, respectively. A regression model was established between the forest volume and the SAR data using a polynomial regression. For the L band JERS-1 SAR data, the forest volume is linearly related to the pixel value when the volume is less than 58m³/ha, then the regression curve gradually flattens and reaches saturation at more than 72 m³/ha. For the C band ERS-1 SAR data, the pixel value has a better linear relationship with the forest volume at less than 40 m³/ha, then the curve flattens and reaches saturation at more than 45 m³/ha. This method has a large potential error for forest volume estimation due to the saturation point. Therefore, the SAR data were used in combination with Landsat TM to establish the multivariate regression model for forest volume estimation over the Luoshan Forestry Centre test site. The error between the predicted and the actual volumes was 2.26%, indicating that this method has potential for accurate volume estimation of forests (Referred to "The Final Report of Spaceborne SAR Applications: Forest Application").

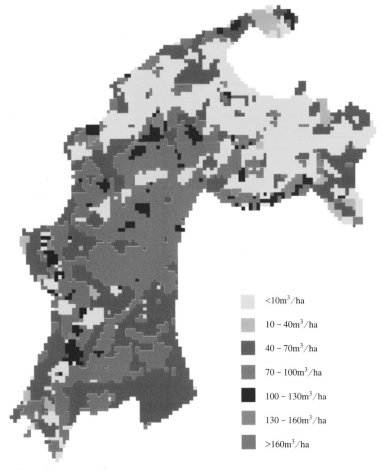

<10m³/ha

10 ~ 40m³/ha

40 ~ 70m³/ha

70 ~ 100m³/ha

100 ~ 130m³/ha

130 ~ 160m³/ha

>160m³/ha

Map of Forest Volume Estimation at Luoshan Forest Centre, Shandong Province of China.

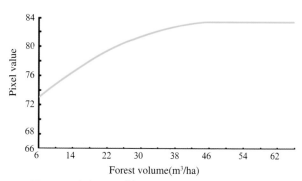

The regression curve of forest volume to ERS-1 SAR data.

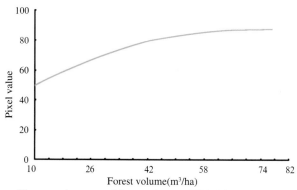

The regression curve of forest volume to JERS-1 SAR data.

HYDROLOGY

For hydrological research with SAR data, many different applications are feasible, especially with multi-band, multi-polarization SAR data.

SAR imagery can 1) clearly display the boundary between water and land, micro-topography, and man-made targets; 2) reveal subsurface features in some situations due to its penetration ability; and 3) provide information on the distribution of vegetation, soil moisture, and surface roughness. These abilities make SAR a good tool for delineating water bodies and studying the migration of rivers, lakes and coastlines.

SAR systems have characteristics valuable for soil moisture detection since the radar response is sensitive to a target's dielectric constant that is mainly determined by water content. For ice and snow research, SAR imaging systems can be used to discriminate different kinds of ice or snow and extract thickness and moisture content of ice or snow quantitatively.

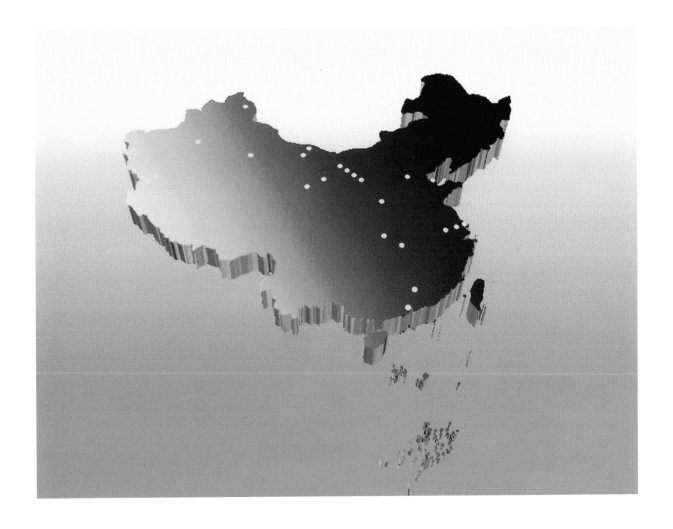

RIVER AND DRAINAGE

Characteristics of River Beds

Yinchuan Segment of the Huanghe River

SIR-C imagery acquired early during the mission shows the 60 kilometer-long Helan-Taole segment of the Huanghe River. West of this image is the northern part of the Yinchuan Plain, and east of the image is the Maowushu Desert. The Yinchuan Plain is one of the most famous irrigated areas in the Huanghe River watershed. About two thousand years ago such well-known irrigation canals as Qinqu, Hanqu and Tanglaiqu were excavated to divert water from the Huanghe River to this region. Now the total irrigated area of the Yinchuan Plain has reached more than 200,000 hectares, and these cultivated lands are mainly devoted to spring wheat and paddy-rice growing.In the SIR-C image, the criss-cross pattern of irrigation canals can be seen. Easily recognized is the Huinong Canal and the fourth drainage ditch that intersects obliquely with the canal. The radar response from a canal or ditch is mainly controlled by its directional orientation to the radar illumination. When a canal or ditch is perpendicular to the radar illumination, HH polarization will give the stronger response. When the radar illumination is at an angle of 45 degrees, HV polarization will give the stronger response. In the false color composite of the SIR-C L-HH, L-HV and C-HV image, cultivated lands display variegated colors that are determined by crop species and growing stage, directionality of the furrows, soil moisture, and other factors.

SIR-C Image of Taole Irrigation Area(R: L-HH, G: L-HV, B: C-HV)

SIR-C Image of Helan area, Ningxia Province (R: L-HH, G: L-HV, B: C-HV)

0   10km

Taole is an irrigated area reclaimed in recent years. During the dry season (when the image was acquired), many channel bars, point bars, and distributaries can be seen, and the wandering of the Huanghe River bed shows up clearly. A buried channel is revealed by obvious tonal differences coming from the surrounding aeolian sand sheets.

portion of the riverbed in this segment is more than 15 kilometers. A large number of channel bars, and point bars are scattered inside the river bed, indicating a typical braided river. The dense bright spots distributed on loess platforms and alluvial plains are towns and villages.

Hancheng-Yongji Segment of the Huanghe River

The Hancheng-Yongji segment of the Huanghe River is located in the south of Fenhe graben, Shanxi Province, north of which is the Yumenkou Gorge, and south of which are the Zhongtiao Mountain. The developing of this river segment is controlled by fault block differentiation inside a Fenhe graben. The loess platform to the south of Linqi is a secondary graben that can be delimited by the loess gullies flowing inside from both the south and north sides. Zhongtiao Mountain is a zone of uplift caused by neotectonic block motion. Two saline lakes (Xiaochi and Yanchi) are located in the piedmont depression. The widest

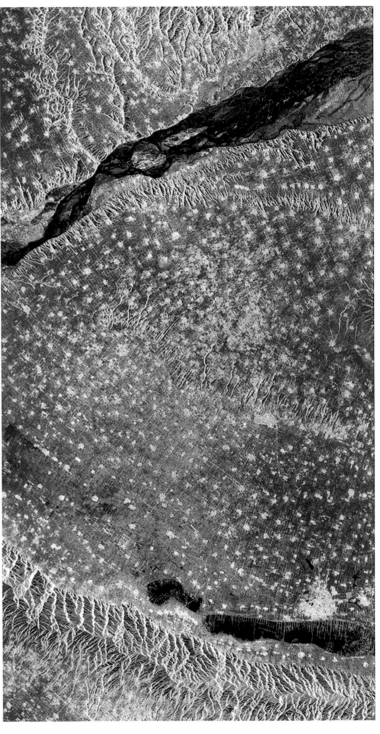

SIR-C Image of Hancheng-Yongji Segment of Huanghe River
(R: L-HH, G: L-HV, B: C-HV)

0   10km

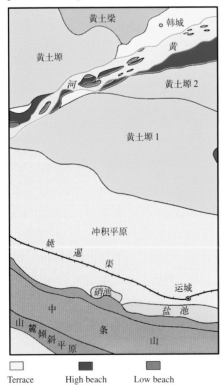

| Terrace | High beach | Low beach |

Landform Interpretation Map of Hancheng-Yongji
Segment of Huanghe River

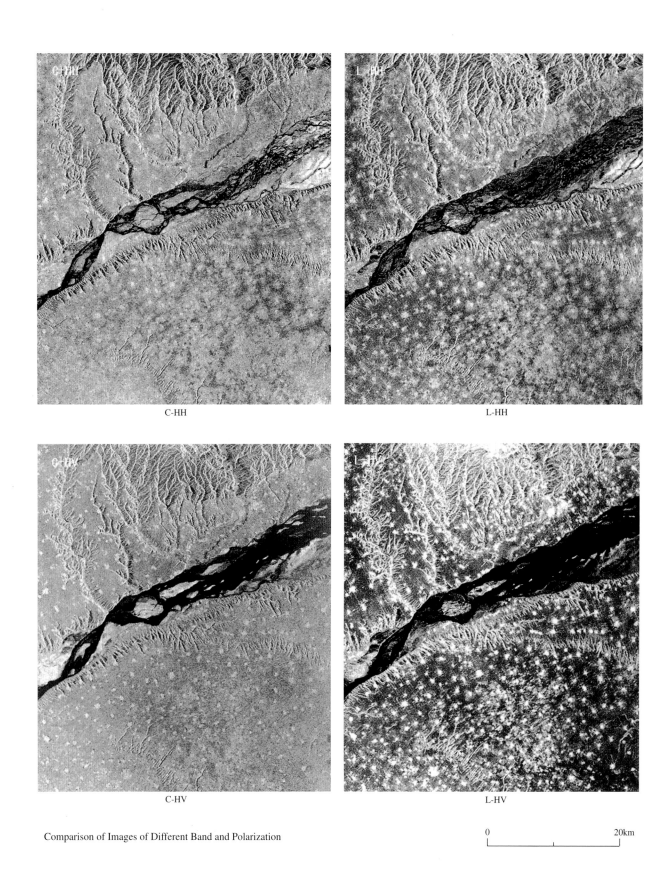

C-HH

L-HH

C-HV

L-HV

Comparison of Images of Different Band and Polarization

0   20km

Middle Course of the Tarim River

The Tarim River rises in the Karakorum Mountains. Its upper course is called the Yarkant River. Downstream from the point where the Aksu River converges with the Yarkant River it is called the Tarim River. This SIR-C image shows one segment of the middle course of the Tarim River that is located to the south of Kuqa. The alluvial plain that extends more than 100 kilometers in a north-south direction was formed by frequent migrations of the Tarim River.

Neotectonic movement controls the course of the Tarim River in a south-north direction. The uplift of the Tianshan Mountains influences the river to shift southward, but in regions of structural stillstand, the uplift of relief caused by the accumulation of alluvial and aeolian layers may force the river to shift northward. This kind of alternating movement left abandoned channels on both the south and north side of the present channel. In the SIR-C image, both the modern channel and the abandoned channel can be seen as meandering streams with oxbow lakes and contorted sinusoidal patterns.

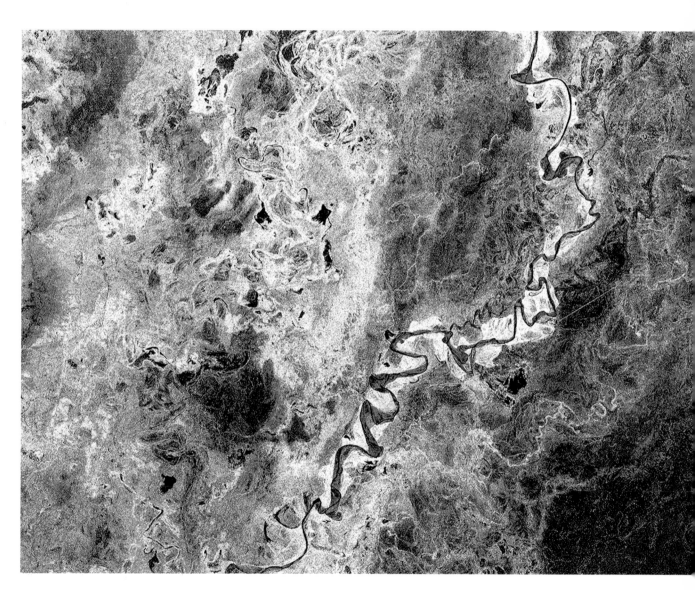

SIR-C Image of the Middle Course of Tarim River to the South of Kuqia City
(R: L-HH, G: L-HV, B:C-HV)

0   10km

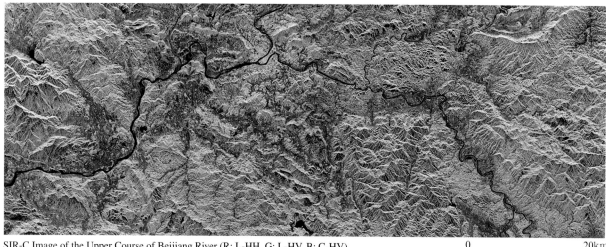

SIR-C Image of the Upper Course of Beijiang River (R: L-HH, G: L-HV, B: C-HV)

0   20km

The upper course of the Beijiang River is a limited meandering stream in a mountainous area. Two tributaries, which are called Wushui and Dianshui, converge at Shaoguan City. From this point downstream, the river is called the Beijiang River. The Dianshui River is the east branch and flows through Cretaceous and Tertiary conglomerates and sandy shales. The limited meandering stream is restricted due to the resistance of mountain lithology. The west branch of the Wushui River flows through a Cretaceous and Permian limestone syncline, and central river bars appear at wider segments in this branch. The Beijiang River has a wider streambed and a higher rate of discharge, and it exhibits a trend to erode eastward due to the injection of western tributaries.

Migration of River Channels

Historical Changes of the Huanghe River Delta

The Huanghe River delta is the modern delta that formed after 1855 when the river breached its dyke at Tongwaxiang and captured the Daqing River. Each year this river delivers several hundred million tons of sand to its delta complex and the nearby coastal area, causing a constantly outward extension of the coast. The littoral zone has a small drainage, and the water depth is less than ten meters. The ratio of stream (fluvial) to tidewater (tidal) reaches 3.6 in the non-high-water season, and can reach 7.5 in the high-water season. The accumulating speed of sand deposition is greater than that of sediment compaction and crustal sinking. These are the dynamic conditions of a rapidly growing delta. In the multi-temporal false color composite RADARSAT image, the intruding delta formation is the newest delta developed since the river flowed into the sea via Qingshuigou.

From 1855 until the present, the river has changed its course many times, demonstrating the ability of the river to deposit silt, breach its levees, and divert its course. What the image demonstrates is the appearance of a complex delta composed of small deltas that have developed at different times. Two centers near Ninghai and Yuwa can be detected through analyzing the distribution of natural levees formed on both sides of present day and abandoned channels. These levees indicate that the shift of the delta apex and the evolution of the estuary has gone through two cycles. Due to the need for the development of the Gudong oil fields, some areas in the intertidal zone have been reclaimed and are now land. Except for the Huanghe River estuary, all beaches are muddy and flat. The north part of the delta is an erosional coast, and the east part is a prograding coast.

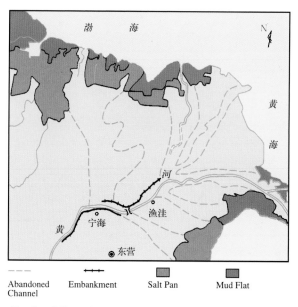

---	+++	▨	▨
Abandoned Channel	Embankment	Salt Pan	Mud Flat

Abandoned Channels and Bottomlands near Huanghe River Estuary

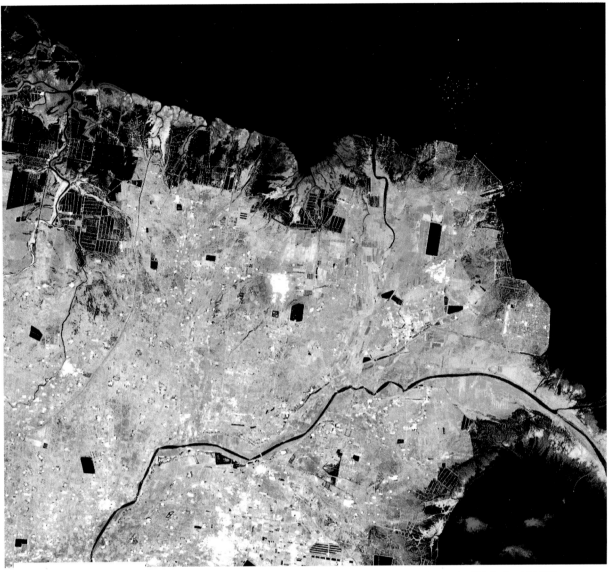

False Color Composite of Multi-temporal Radarsat Images of Huanghe River Estuary (R: Jul.9 1997, G: Aug.2 1997, B: Sep.10 1997)

Change of the Yangtze River Estuary from 1976-1998

The Yangtze River delta is a compound delta, the apex of which is located in the Zhenjiang-Yangzhou area. There is a big lagoonal swamp plain outside each bank. Taihu is in the center of the northern plain, and Gaoyouhu is in the center of southern plain. The Yangtze River delivers some 500 million tons of silt and sand into the sea each year through its branching estuary, forming many sandbars, of which Chongming Island is the biggest. Due to frequent swings of the main stream channel and the migration of outlets, the geomorphology of the Yangtze River estuary is constantly changing. Changes in the estuary in the past 20 years can be delineated by comparing the RADARSAT ScanSAR image acquired in 1998 with the registered Landsat MSS image acquired in 1976.

The most obvious change on Chongming Island is

that its area has greatly extended. Especially in the eastern part, the newly reclaimed farmlands exceed the scope of the original high flat. Two stages of reclamation can be detected, and an enclosing dam, the dividing line between two stages of reclamation, can be discerned. Intensive silting occurred in the branch to the north of Chongming Island, some new sandbars came into being, and the north bank of Chongming Island and the south bank of Qidong were also extended.

The high tidal flats in the estuary have been changing constantly. Some have been reclaimed to farmlands, some have disappeared, and some have linked to each other. Changxing Island and Hengmen Island are becoming connected together. Although ScanSAR data have a lower spatial resolution, they can still be used to illustrate the land/water boundary, villages and cities, and the structure of farmlands. A good example is that sandy fields with a pattern of paddy fields alternating with ponds can be distinguished.

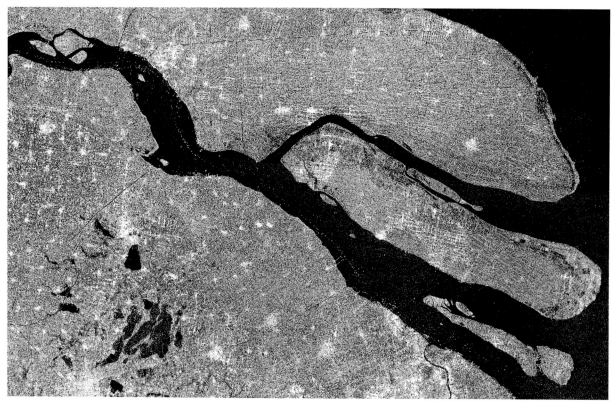

Radarsat ScanSAR image of Yangtze River Estuary

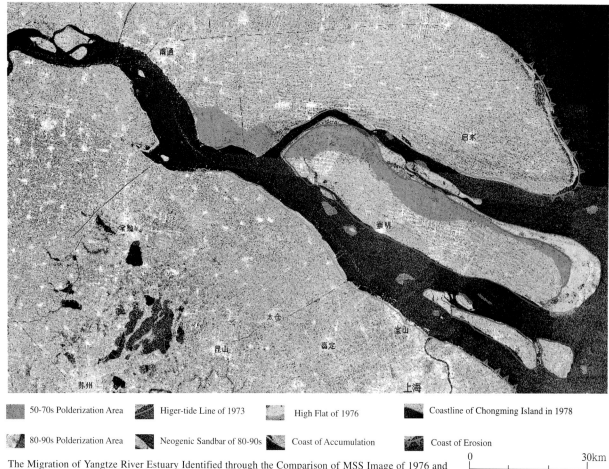

| | 50-70s Polderization Area | | Higer-tide Line of 1973 | | High Flat of 1976 | | Coastline of Chongming Island in 1978 |
| | 80-90s Polderization Area | | Neogenic Sandbar of 80-90s | | Coast of Accumulation | | Coast of Erosion |

The Migration of Yangtze River Estuary Identified through the Comparison of MSS Image of 1976 and Radarsat ScanSAR Image of 1998.

0   30km

Migration of the Ejin River

The ScanSAR data of 50 meter spatial resolution acquired in December, 1996 covers the entire Ejina great alluvial fan and the BadanJaran Desert. The Heihe River originates in the north piedmont of the Qilian Mountains. Its lower course is called the Ejin River (Ruoshui in ancient times) and it is one of China's largest inland drainage systems. It divides into two branches at Huxixincun; the east branch flows into Sogo Lake, and the west branch flows into Gaxun Lake. Presently, the Ejin River flows via the east branch into Sogo Lake that has a surface area of some 20 square kilometers. It can be seen that the river channel is fairly narrow, presenting a dark tone in the image. Gaxun Lake has fully dried up, presenting a circular bright tone caused by remnant dykes of an ancient lake. The bright tone on both sides of the main stream channel of the west branch and in the reclamation area near the town of Ejin is due to higher soil moisture caused by the enrichment of groundwater and the subsequent growth of vegetation. Some abandoned channels still have flat, silty-sandy surfaces, which present a dark tone.

Historically, migration of the Ejin drainage channel has occurred often. In 1959 ~ 1969, it flowed via two branches, and Gaxun Lake still had a surface water area of 255 square kilometers. In a Landsat MSS image acquired in 1972, it could be interpreted that the Ejina

Radarsat ScanSAR Image of Ejin County

0   40km

63

River flowed via its east branch. At this time, Sogo Lake had a surface area of 60 square kilometers, but the surface area of Gaxun Lake had been reduced to 105 square kilometers. Swan Lake lies to the east of Sogo Lake, and now has only a small surface water area, but it is evident Swan Lake had quite a large water area in the past. It is recorded in historical documents that in the Jin Dynasty, the Southern Dynasties, and the Northern Dynasties (317 ~ 534 AD), the Ruoshui river flowed into Swan Lake which at that time was called the "Juyan Lake." In the Northern Song Dynasty (907 ~ 1234 AD), the Ruoshui River flowed via two branches into Gaxun Lake and Sogo Lake, at which time the two lakes were connected. In the Yuan Dynasty (1206 ~ 1368 AD), the Ruoshui river again flowed into the "Juyan Lake" (location of current Swan lake). From the Ming Dynasty (circa 1368) on, the "Juyan Lake" began to shrink. The Radar image reveals two abandoned channels that connect the east branch of the Ejin River and Swan Lake, and some older abandoned channels to the west of the west branch of the Ejin River also can be discerned.

Detection of Abandoned Channels

An abandoned channel is an old channel left after river diversion. An abandoned channel is usually located in some kind negative landform (e.g., low topography) and has special soil and water content conditions. Since radar has the ability to detect water-land boundaries, micro-relief, soil moisture, and vegetation distribution, it can be very useful for delineating abandoned channel. The significance of studying abandoned channels is to provide information relevant for consolidating dams and developing anti-flood policies.

The Migration of the Jingjiang River Channel

Numerous abandoned channels formed during different times in the Jingjiang watershed. Since abandoned channels have special micro-geomorphologic and hydro-geological features, radar images can reveal their locations and shapes. By relying on the clues of ancient river channels provided by historical documents combined with traces of abandoned channels interpreted from radar image, the evolutionary history of the Jingjiang watershed can be roughly recovered.

Most of the abandoned channels detected on the radar image coincide with the clues provided by historical documents. One important rule for determining the active stage of an abandoned channel is to realize that the river channel is relatively stable between big floods which are main factors causing the shift and abandonment of a river channel. Through integrated analysis of the image and interpretation of historical documents, a conclusion can be reached that before the Ming Dynasty, the Jingjiang River channel was stable and had numerous branches on both banks, so that floodwaters were easily discharged. Following

L-SAR Image of Jingjiang River Acquired on Jul. 27-31 1998

0   20km

the intensifying of human reclamation and the shrinking of lakes, the formation of river meanders and diversion channels increased. In early times, the meanders first appeared in the lower course, then gradually extended to the upper course. However, in recent centuries the meanders appear along the entire course; this is related to the silting of Dongting Lake which caused lake waters to fill up the river channels.

There are twelve episodes of meander cutoff recorded in historical documents, but the radar imagery reveals that cutoffs even occurred at Xiaozoujia, Jianglaowan and Yazihu, and the occurrence of these cutoffs was obviously caused by a chain reaction. This situation can occur when several meanders have a radius of curvature higher than 2.5 and floodwaters from an especially large flood breaches a meander neck. An initial meander cutoff leads to a subsequent chain reaction creating a series of downstream cutoffs. In recent decades, meander changes have been fairly small; the total length of the river has increased only 1.6 kilometers and the average radius of curvature of meanders has increased by only 0.02. These

facts document that the development of river meanders has been brought under control and the channel modifications of this river are effective.

The Migration of River Channels and Flood Protection Dyke Section Analysis

Through a comprehensive and integrated analysis of GlobeSAR data of 6 meters spatial resolution, RADARSAT standard mode data of 25 meters spatial resolution, fine-mode data of 10 meters spatial resolution, and historical documents, nine abandoned channels were detected in the West River and North River watersheds. Abandoned channels A, B, G, H and I, recorded accurately in historical documents, are easily discerned in the radar image. Abandoned channels C, D, E and F are known to have come into being 2000 years ago by historical documents, but no detailed information about their location and shape is recorded. Radar images reveal not only the position of these abandoned channels, but also their intersection relationship.

Migration of a river channel results from watercourse blockage that leads to diversion of the old channel. This can be caused by both natural agents and human activities. One thousand years ago, the North River was one of the tributaries of the West River, but a blockage caused the West River channel to swing from northeast to south and to flow into the sea separately. The North River flowed into the sea via the abandoned West River channel. The L-shape of the river channel near Qingqi is verification of this point of view.

There are four kinds of potentially dangerous dyke sections which are prone to resist flooding: (1) meanders of a river course, especially near a concave bank with a large radius of curvature; (2) outlets of narrow river sections; (3) confluence points of river channels; and (4) dykes or dams that exist where an abandoned channel crosses underneath. Through analysis of dyke and dam systems in this area, seven potentially dangerous sections were delineated. The total length is some 69.4 kilometers, the area of affected farmland is about 46,500 hectares, and the affected population is about 510,000.

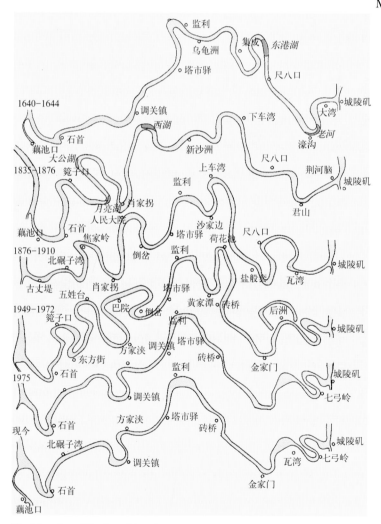

The Migration of Jingjiang River Channel from 1640 A.D. to Current Time

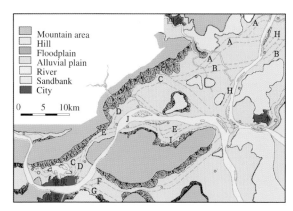

Legend:
- Mountain area
- Hill
- Floodplain
- Alluvial plain
- River
- Sandbank
- City

0 5 10km

A. Inferred pre-historical channel
B. Inferred pre-historical channel
C. Channel in 475 B. C. to 221 B. C.
D. Channel in 206 B. C. to 220 A. D.
E. Channel in 220 A. D. to 420 A. D.
F. Channel in 618 A. D. to 970 A. D.
G. Channel in 960 A. D. to 970 A. D.
H. Channel in 1368 A. D. to 1644 A. D.
I. Channel in 1644 A. D. to 1911 A. D.
J. Current channel

The Distribution of Abandoned Channels in Zhaoqing Area

The information of regions prone to suffer flood

Region	Length of embankment (km)	Length of dangerous dyke (km)	Area of cultivated land (ha)	Population (10,000 P)
A	3.5	4.5	3340	5.82
B	6.0	6.0	10600	13.33
C	1.40	9.0	2980	3.49
D	17.0	7.0	8500	7.36
E	18.0	5.0	6500	4.81
F	10.4	10.4	1180	1.01
G	44.0	27.5	13387	15.21

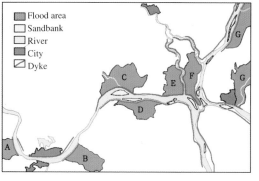

Legend:
- Flood area
- Sandbank
- River
- City
- Dyke

The distribution of anti-flood dangerous dykes in Zhaoqing area

3D perspective view of Zhaoqing area (Radarsat image draped on DEM)

Composite of multi-temporal Radarsat images of Zhaoqing area

0   20km

Water Conservancy Projects Along the Sanxia Gorge of the Yangtze River

The Sanxia Gorge hydro-junction project, which commenced on December 24, 1994, is now the biggest hydroelectric project in the world. Its building progress from the end of 1996, when the diversion canal was begun, to November 1997, when the second-term cofferdam project was being finished, has been recorded on multi-temporal Radarsat images. The longitudinal cofferdam was built on the basis of a central bar, Zhongbao Island. The diversion canal indicates the watercourse between this island and the south bank of the Yangtze River, which needed to be deepened, and the silt, which need to be cleared. The upstream cofferdam and the downstream cofferdams were completely built by casting earth and stone.

On the image acquired on December 16, 1996, the progress of the diversion canal project can be detected. The diversion canal was being dug and the east and west sides of the building site had been enclosed by a temporary dam. The area of the building site was about 97 hectares. The construction of this diversion canal was such a giant project that it was called the "man-made Yangtze River." The newly built highway bridge and some thirty ships are especially visible on the image due to their high radar return (high brightness value).

On the image of September 8, 1997, it can be seen that the second-term cofferdam project has started. Four intruding heads of the upstream cofferdam and downstream cofferdam have appeared, and water has filled the diversion canal. The watercourse between the temporary ship lock and the permanent ship lock was being dug, and the embankments of both banks to the east of the second-term cofferdam construction site had been built up.

On the image of October 7, 1997, new progress on the Sanxia Gorge project can be detected. There was only a little extension of the four intruding heads of the upstream cofferdam and the downstream cofferdam due to the delay of the high-water season. The width of the upstream cofferdam was about 460 meters, and that of the downstream cofferdam was 480 about meters. October 7th was the first day for the diversion canal to be open to navigation. One month later, work was finally finished.

The image in the Yichang area shows the full view of the Gezhouba hydrology hub. Many facilities, such as ship locks, flood gates, power stations, protection embankments, and switching stations can be detected.

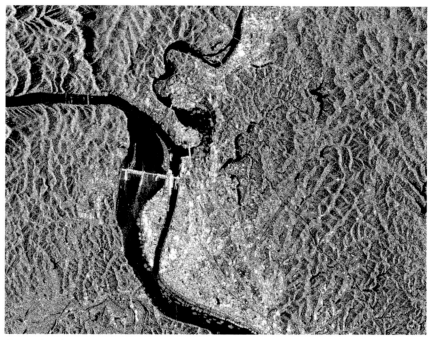

Composite of multi-temporal Radarsat images of
Gezhouba Hydrology hub

0   5km

Composite of multi-temporal Radarsat images of Sanxia Gorge area
(R: Dec. 16 1996, G: Oct. 7 1997, B: Sep. 8 1997)

0   10km

Composite of multi-temporal Radarsat images of Sanxia Gorge hydro-junction project building site

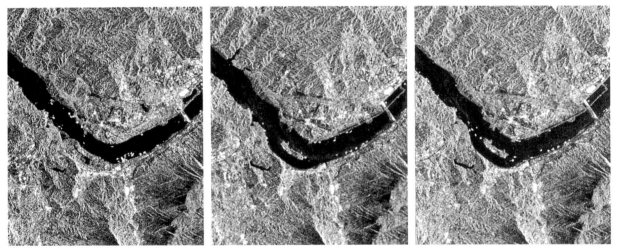

Comparison of Radarsat images acquired at different date (Left: Dec. 16 1996, Mid: Sep. 8 1997, Right: Oct. 7 1997)

L-SAR image of Sanxia Gorge hydro-junction project building site

0   5km

LAKES

Limnological Environments

Taihu Lake

Taihu Lake, one of China's five largest freshwater lakes, is located at a depression in the south of the Yangtze River delta. Its west bank is adjacent to rolling hills which are a branch of the Tianmu Mountains, and its east bank is 100 kilometers from the East China Sea. Taihu Lake is a crescent-shaped, large, shallow lake. It has a water area of about 2,425 square kilometers, and its average depth is only about 2.12 meters. Its circular west bank is a slightly uplifted accumulation bank, and there exist lower-grade terraces on the piedmonts of nearby mountains and sides of river valleys that flow into the lake. The east and north bank, characteristic of meandering and numerous bays, is a recessional eroded bank in a state of slow depression, and lake water is gradually intruding toward the land, carving cliffs of 2 to 3 meters high along lake-facing banks of many peninsulas.

There are a great number of islands inside Taihu Lake. Historical documents reveal that there have been seventy-two islands recorded in Taihu Lake, but only

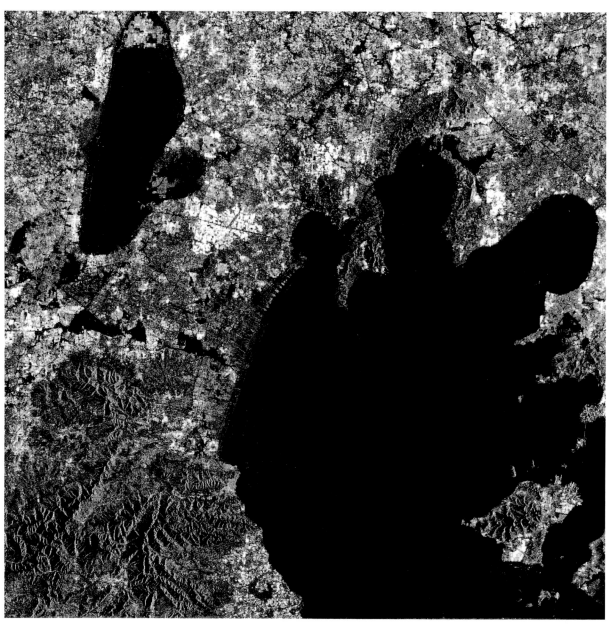

JERS-1 SAR image of Taihu Lake

0   20km

forty-eight were found to be present by a 1960 investigation. Since then, Majishan, Chongshan and Baifushan were three islands reclaimed, and they became peninsulas connecting land. Most of the islands are rocky, are scattered along a north-south direction, and are concentrated near Dongting West Mount Island. Dongting West Mount Island is the largest island, with an area of 62.5 square kilometers. Piaomiao Peak on this island is the highest point within the Taihu Lake region. There are forty-two submerged reefs in the lake, most of which are located in west Taihu Lake. These submerged reefs are dangerous to shipping.

Lakes in Xuanzhou-Liyang Area

The X-SAR image acquired in April 1994 covers the area from Xuanzhou, Anhui Province northeastward to Liyang, Jiangsu Province. The X-band radar is extremely sensitive to any slight undulation of a target's surface. This leads to a low contrast between land targets, but an obvious high contrast between land and water, or between a pure water surface and a vegetation-growing water surface. Thus X-band imagery is useful for delineation of water bodies.

The left half of the image shows rolling hills and lakes of different sizes. The big lake on the upper-left of the image is Nanyi Lake. There are two rivers that flow into the lake from the east whose watercourses are now confined by man-made dykes. The low-lying land between the two rivers has long been a floodplain. The existence of numerous curved, beheaded rivers provides proof. This low-lying land is a high-yielding farming area and also a key area for preventing flood. Nanyi Lake is an important flood storage lake in this area, and the non-utilized bottomlands and dams in the south and east sides of Nanyi Lake can easily be discriminated. The large lake on the upper-right of the image is Changdang Lake. The former marshland to the south of this lake has been reclaimed for farmland and fishponds. The right half of the image shows densely scattered lakes, rivers, canals, drainage ditches and farmlands. This kind of scene is typical of the area to the south of the Yangtze River in China.

Mountain	Plain	Lakeshore	Hill	Water body	Reclamation land

Landform interpretation map of Taihu Lake nearby area

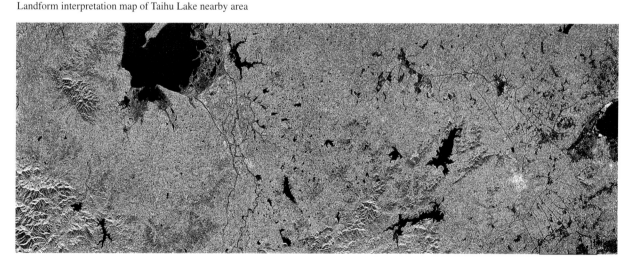

X-SAR image of Xuanzhou-Liyang Area

Qinghai Lake

Qinghai Lake is the largest interior plateau lake in China. Its formation dates to the Pliocene epoch. This lake is joined to the surrounding mountains by a series of faults. It is the water-collecting ground for a large arid area, and there are two rivers flowing into it. Due to no outlets and intensive evaporation, the lake has become brackish. In 1981, the altitude of its water surface was 3,913 meters, its surface water area reached 4,340 square kilometers, the calculated water depth was 27 meters, and the total water-storage capacity was approximately 77.8 billion cubic meters. The lake has an obvious weather-adjusting effect on its surrounding environment, leading to a higher rainfall.

The false color multi-band, multi-polarized SIR-C image reveals a great number of remnant ancient lake strandlines to the south of the lake. There usually exists a terrace of 50 ~ 60 meters above the current strandline. The surface of this terrace is covered with alluvial fans stemming from the north piedmont of the South Mountain, Qinghai. The Qinghai-Tibet highway was built along the front of this terrace. Lakeshore gravel layers and lacustrine fine sands are found in boring logs. This indicates that the terrace was of lacustrine origin, formed during the recession of the lake. Due to the cover of alluvial fans, only a small remnant of the terrace near Erlangjian remains. Subsequent to the climate becoming more arid during the Holocene epoch, the lake began to shrink. In the last three decades, the average annual shrinkage rate reached 10 centimeters/year. This has caused damage to the ecological balance.

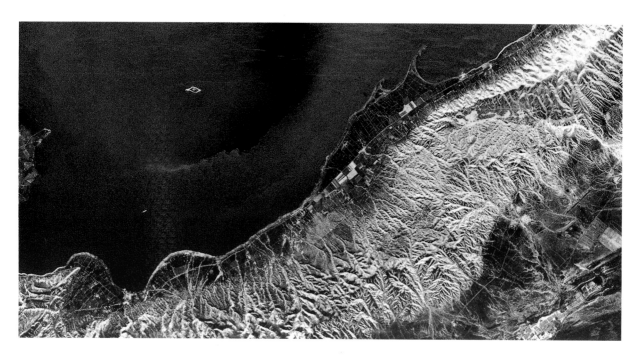

SIR-C image of southern part of Qinghai lake (R: L-HH, G: L-HV, B: C-HV)

0   20km

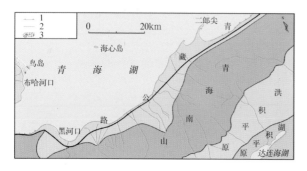

Interpretation map of Lakeshore landform

1. Lakeshore of erosion

2. Lakeshore of accumulation

3. Shallow beach

MSS image of southern part of Qinghai lake

Evolution of Lakes

Wulungu Lake

Wulungu Lake is located in the north of Xinjiang Autonomous Region, north of which is the Altai Mountain range and south of which is the Junggar Basin. The small lake to the southeast of Wulungu Lake is called Jili Lake. There the Erqisi River flows north of Wulungu Lake, and the Wulungu River flows into the two lakes from the southeast. Due to its VV polarization and small incidence angle (23°), ERS-1 is suitable for detecting water state, that is the roughness and undulating status of the water surface. Thus for most cases, the water body yields varying tones related to water state on an ERS-1 Image. The image shows an obvious gray-level difference inside Wulungu Lake; the northeastern part is brighter than the southwestern part. It can be inferred that air currents causing different wave climates on the water surface bring about this phenomenon. The west

bank of the lake is an erosional bank. The east bank of the lake is a prograding bank. Under the influence of the prevailing west winds, the lake current transports bottom sediment to the east bank forming sandbars.

The image reveals that Wulungu Lake and Jili Lake had been connected in the past. Following the north-to-south down-stream swing of the Wulungu River, the continual accumulation of sediment from this river caused blockage of the area between the two lakes. The delta area between the two lakes has been reclaimed for farmland. To the northeast of the lake, there are several small lakes. The cause of their formation can be traced to sandbars obstructing water to form "lagoons." From the Holocene Epoch on, the climate in northwestern China became more and more arid, and shrinkage of the lakes was accelerated. The radar image provides direct evidence for this trend.

ERS-1 SAR image of Wulungu Lake in the north of Xinjiang Autonomous Region

0   40km

Aksayqin Lake

Aksayqin Lake, near the boundary of Xinjiang and Tibet, is situated on the Qinghai-Tibet Plateau of western China. The lake was formed in the fault basins of southern side of the Kunlun Mountains. The fault basins, passing several lakes in the Qinghai-Tibet plateau, consist of a number of faults and valleys oriented in EW. The lake primarily originated from a group of Kunlun glaciers which are elevated at more than 6,000 m. There are totally 3180 glaciers, in which more than ten mountain valley glaciers are developed. The Aksayqin River, a main water supplier to the Aksayqin Lake, was originated from the melted water of these glaciers. In the Aksayqin River catchments, there are 129 glaciers. The total glacier area is 709.08 km^2, and total ice reserves is 136.2698km^3. The Litian, Keqikebing, and Qiongbingshui rivers input water to the Aksayqin River. The total length of Aksayqin river is 110 km, and total catchment area is about 5670 km^2. According to previous studies, this area was connected with the Tiansuihai area to the northwest, allowing for the formation of a huge paleo-lake with an area of approximately 1,400 km^2 and with the highest shoreline at 4,860 m. The measured area of Aksayqin Lake derived from the X-SAR image is 168 km^2 indicating the lake has retrograded extensively. The many sand barriers around Aksayqin Lake also provide evidence for the presence of paleo-lake retrogradation.

Aksayqin Lake is a saline lake; the current surface area is 4,840 m above sea level. According to previous research, it is common to see 8 to 10 water retrogradation rhythms in the paleo-lakes of the Qinghai-Tibet Plateau. There are eight retrogradation rhythms at Aksayqin Lake, indicating there were eight dry seasons since the last interglaciation.

According to previously reported ^{14}C dates for lacustrine sediments, the evolution of paleo-lakes in the western Kunlun Mountains can be classified into three phases: 1) *Circa* 46,000 years ago the Tiansuihai paleo-lake had an area of 3000 km^2. At this time, the altitude of the uplifted Qinghai-Tibet Plateau could not completely prevent the entrance of warm and wet air currents from the Indian Ocean. A subsequent dry interval led to the disappearance of this paleo-lake. 2) *Circa* 35,000 years ago, several individual lakes were connected together related to the warm climate interval correlated with retreating glaciers. 3) *Circa* 18,000 years ago, the very extensive paleo-lake of Aksayqin-Tiansuihai was formed. Subsequently, after entering the Holocene, the climate became warm and dry, and the huge paleo-lake gradually receded.

The SIR-C false color composite image (R: L-HH, G: C-HH, B: X-VV) shows Aksayqin Lake and river courses from which several old shorelines can be identified. Comparison of single band and various polarized methods suggests that L-HH is best for recognizing shoreline features.

L-HH(R) C-HH(G) X-VV(B)

SIR-C image of Aksayqin Lake

0   8km

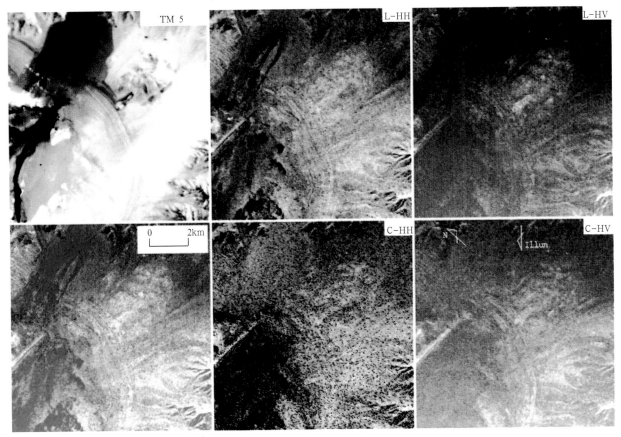

Comparison of various images of Tianshuihai ancient lake showing relict lakeshore line.

X-SAR image of Aksayqin Lake

Lop Nur

The Lop Nur is located east of the Tarim Basin. On the RADARSAT image, the ear-shaped ancient lake basin is extremely visible. The N-S trending, strip-like new lake is located to the west of the ancient lake. From the north it receives water from the Kongqi River, and to its south is the terminus of the Tarim River. The form of the Lop Nur depression is controlled by two groups of faults oriented NNE and NE. Northeast of the intersection point of the two groups of faults is the rhombus-shaped North Mountain, and southwest of that is the Lop Nur depression where a large ancient lake once existed. Image analysis reveals the evolution of the Lop Nur. The flexural uplift of the north part of the ancient lake basin influenced the lake water to shift southward. In the north part the ancient lakeshore dykes are arranged in a continuous ring-shape, but in the south part the ancient lakeshore dykes are overlapping, helping to verify this inference. The westward shrinkage of the ancient lakeshore dykes is a reflection of a NNE block movement. Following this shrinkage, the delta of the Tarim River constantly migrated westward until the position of the new lake was reached.

The water area of the ancient lake was fairly large, as indicated by traces of an ancient lake basin. Climate change was the main factor causing shrinkage of the ancient lake, but frequent river diversion was also and important factor. In 1921, the Tarim River was diverted to flow into the Lop Nur via the Kongqi River course. In 1952, a dam was built across the Tarim River, causing it to flow into Taitema Lake. In the late 1950's and afterwards, a number of reservoirs and dams were built on the Kongqi River and the lower reaches of the Tarim River in order to divert water for irrigation. As its water supply was thus cut off, the Lop Nur gradually dried up.

On the SIR-A image, the ancient lake strandlines of

Standard mode Radarsat image of Lop Nur

0   20km

the Lop Nur are also obviously visible. Aeolian accumulation, wind erosion features, and saline lake landforms are distributed in the ancient lake basin. For example:

1) Yardang, an erosional wind feature, is evident at H5-J1. The length of this land feature is about 22 kilometers. This yardang was formed at the base of a dried ancient lakebed. The salt crust was eroded to countless long mounds and valleys of different lengths and widths ranging along a NE direction, the direction of the prevailing winds. This special landform yields alternating bright and dark tones on the image.

2) A salt lake depositional plain, evident in the center of the image, is the dried Lop Nur lakebed of modern times. The long axis of this feature is in a N-S direction and short axis is in the E-W direction. Wind erosion has not yet transformed it to yardang. The tonal difference here results from substantial hydrological differences.

3) Dried salt crust, evident at F1-F2 and H1, is due to the high surface roughness and dielectric constant of salt, which yields a bright tone on the image.

4) Sand accumulation areas are evident at F3 (flat thin-sanded land), G2-3/H2-3 (flat thick-sanded land), and G2/H2 (dune chain).

5) The modern outline of the ancestral water lines, evident at D5-H5, reveals the lake strandlines over time, indicating that the shrinkage trend of the Lop Nur was from northeast to southwest.

Outside of the Lop Nur depression, several other types of landforms can be discriminated. For example:

1) Abandoned channels of the Kongqi River, evident on the left of the image, once formed a dense braid-like drainage that flowed into the Lop Nur at earlier times. Relics of the ancient city of Loulan show as a bright speckle at A4.

2) The east edge of the Lop Nur desert, evident at A4 as a mottled gray and black tone, is the gravel-covered Gobi Desert.

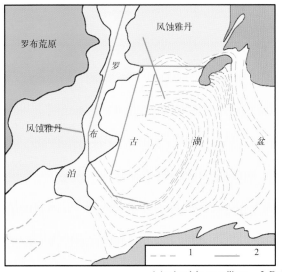

1. Ancient lake strandlines   2. Fault

Interpretation map of Lop Nur relict lake basin and eolian landforms

TM image of Lop Nur area

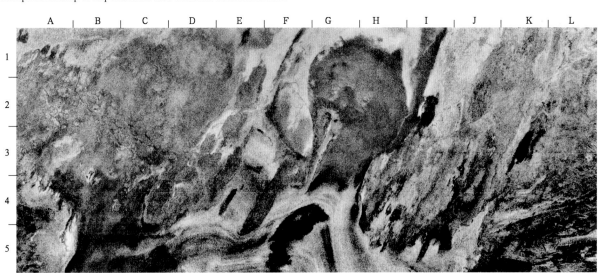

SIR-A image of Lop Nur area

3) A dried pluvial plain, evident at D1/D2-E2, is the front edge of the Kuluketage alluvial fan.

4) Wind erosion valleys and unaka, evident at L1-K5/L5, derive from a bedrock-exposed low-hill area subject to intensive wind erosion and mechanical weathering.

Hongzhe Lake

The image below covering part of Hongzhe Lake was acquired on July 1, 1994 by the X-band SAR system developed by the Institute of Electronics, Chinese Academy of Sciences.

Due to the sensitivity of the X-band radars to even slight undulation of the ground surface, the various targets reveal a weak contrast on the image. However, for some situations, this characteristic of the X-band is useful for target detection. For example, the networks of ropes and buoys and boundary markers delineating aquaculture operations are visible as bright tones contrasting the dark toned water surface. Fin and/or shellfish operations are visible on either side of the delta and in several lakes adjacent to the river.

Comparing the 1994 SAR image with topographic maps surveyed in the 1970s, it can be seen that several new channel bars have been formed from the estuary of the Huaihe River to Hongzhe Lake. These channel bars have divided the river into two narrow branches. This is probably testimony to the shrinking of this lake, but may also be the result of sediment deposition and delta growth.

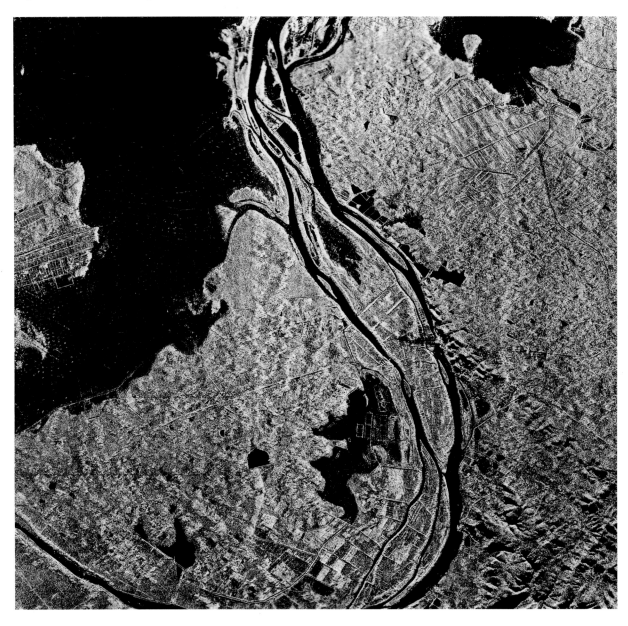

CAS/SAR image of the southern part of Hongzhe Lake

Saline Lake

Saline lake and relevant landform types are easy to be discriminated on multi-band, multi-polarization SAR image. Jilantai salt lake is situated in an asymmetric down-warped basin. The former research reported: during Tertiary period, the lake basin was fairly large, and present Wulanbuhe desert grew on the basis of this lake basin. Controlled by activation of NNE fault and climate change, the shrinkage of this salt lake was accompanied by the substantial differentiation and chemical depositional differentiation. From edge to center, emerges a gravel-sand-sludge gradually varied succession. The majority of chemical sediment in present salt lake is sodium chloride, which is the last differentiated chemical substance. This proves present Jilantai salt lake

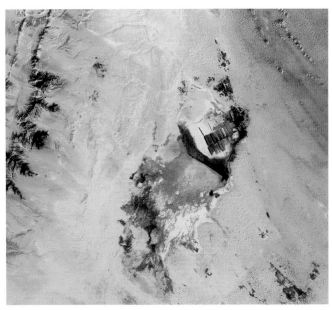

Landsat TM Image of Jilantai Salt Lake in Alasan Plateau

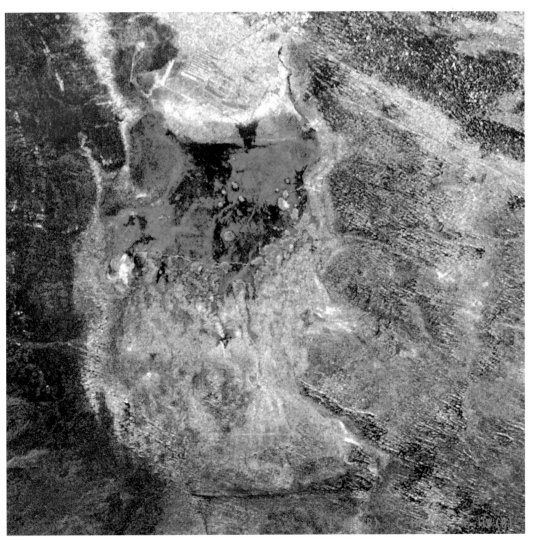

SIR-C Image of Jilantai Salt Lake (R: L-HH, G: C-HH, B: C-HV)

0　　　　　　　　　5km

is the depressing center in this area. An ancient lake strandline is detected on SIR-C image, which is the sand-covered lakeshore terrace front steep bank. It can be inferred that during the latest high-water-level period, the lake basin is at least one times larger than present lake basin. On SIR-C image, saline land yields an obviously brighter tone than desert plain and Gobi flat, due to its high dielectric constant caused by saline itself and higher soil moisture, or undulating small-dune surface formed by plants obstructing sand. The color variation between saline lands on SIR-C image false color composite image is mainly caused by different surface undulating size, as proved by field investigation. The white region indicates the surface of various size undulation, the pink region indicates the surface of tens centimeter undulation, the cyan region indicates the surface of several centimeter undulation, and dark green region indicates flat wet salina. This is only a rough conclusion, but the rule that saline land can yield a stronger radar return than non-saline desert plain and Gobi flat is also verified by images of Gouchi salt lake and Huamachi salt lake. These two lakes locate in adjacent area of Yanchi county, Ningxia province and Dingbian county, Shanxi province. They all still have bigger water area, and the even salt pads are visible on the images.

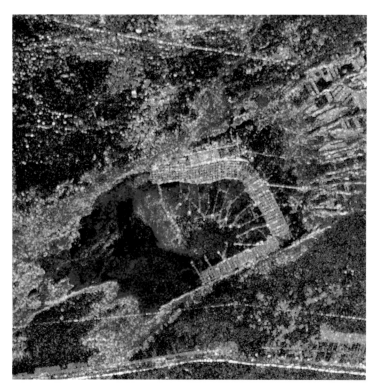

SIR-C Image of Huamachi Salt Lake in Shaanxi Province (R: L-HH, G: L-HV, B: C-HH)

0   1km

SIR-C Image of Gouchi Salt Lake in Shaanxi Province (R: L-HH, G: L-HV, B: C-HH)
0   1km

Jilantai Saltworks (Upper: Salt Pan, Lower: Alkali Flat)

GROUND WATER AND SOIL MOISTURE

Ground Water Storage Condition in Bayinnuorgong Area

Multi-band, multi-polarization radar images can provide soil moisture and surface roughness information and in some circumstances the multi-dimensional data can reveal subsurface information. Therefore, this kind of remotely sensed data has high potential for detecting ground water storage conditions.

The imaged area is Bayinnuorgong in the west of Inner Mongolia. The red line indicates interpreted buried faults. The area marked "A" is a ground water enrichment area, a buried channel or buried lake. Higher water content of the soil and salinization make this kind of area yield strong backscatter response from L-band radar. The surface of "D" is similar to the "A" area, which is characteristic of gilgai with a diameter of approximately one meter. The areas marked "B" are undulating parts of the Gobi Desert. They have a strong

SIR-C image of Bayinnuorgong area (R: L-HH, G: L-HV, B: C-HV)

0   10km

81

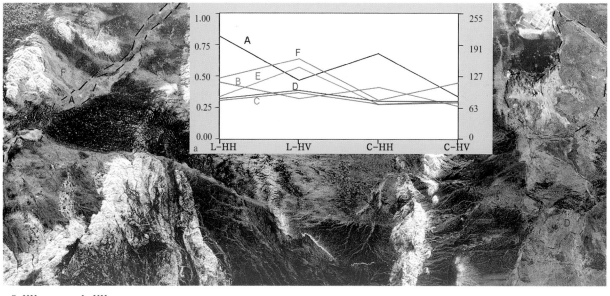

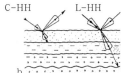

20~30-meter-thick dry sand layer(Water content is near 0%)

20~30-meter-thick dry sand layer(Water content is near 0%)

15~20-meter-thick wet sand and pebble layer
groundwater level

A. salting straw-dune land   B. Straw-dune land   C. dry riverbed

D. Groundwater enrichment region E&F bedrock covered by eolian sand

0   20km

a. Grey-level features of different targets on images of different band and polarization
b. Sketch map of radar wave penetrating dry sandy riverbed

The comparison of radar echo features of different landform units.

response for C-band radar as can be explained by its roughness characteristics. The areas marked "C" are bedrock covered by a sand layer that is no thicker than one meter. They have a different response signature than typical desert, where the thickness of sand is more than several meters. Radar waves can penetrate the thin sand layer of dry sand to interact with the bedrock surface and still yield a strong radar response. The area marked "D" is the alluvial or pluvial plain with small dunes formed around obstructing straw or bush on the aeolian sand. Areas marked "E" and "F" are two depressions with thick, loose sediment. They have a smooth surface, and almost no vegetation due to high permeability of the subsurface substance and low rainfall amounts.

Some of the river channels have a sandy smooth surface that produces a mirror-like reflection surface for both the L- and C-band radar, but on occasions they can still yield a stronger response from the L-band radar due to penetration. From the profile illustration, it can be seen that this occurs when there is a wet mixed sand and pebble layer under several tens of centimeters of sand. It is possible that the stronger response for L-band radar comes from the interaction between the radar wave and this layer.

Ground Water Enrichment Belt in Ejina County

A vivid example of using radar to detect ground water is revealed by the SIR-C image of the Ejina area, Inner Mongolia. Almost all the gullies and abandoned channels are visible on this image. This is not the case on the optical remote sensing images. Dense small dunes are formed around obstacles such as straw or bush which obstructed aeolian sand from scattering in these gullies or abandoned channels. The optimum size and shape of these dunes causes the strong response to the L-band radar.

There is a special fog-like green belt in the middle of the image. Field investigations indicated that this belt is a ground water enrichment area formed along a fault. The clay layer of the southern uplifted side of the reverse fault obstructed the ground water movement which percolates from the N-S tilting piedmont plain. The unique high response characteristics of L-HV can be explained by the high water content of vegetation and soil causing strong volume scattering.

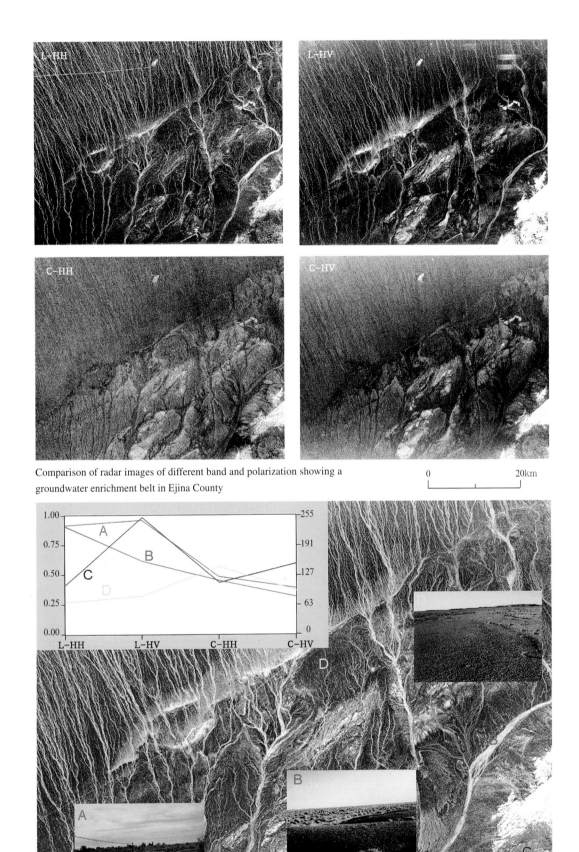

Comparison of radar images of different band and polarization showing a groundwater enrichment belt in Ejina County

0   20km

Comparison of radar echo features of several targets   A. Groundwater enrichment belt B. Dry riverbed
                                                        C. Granite body D. Gobi desert

0   10km

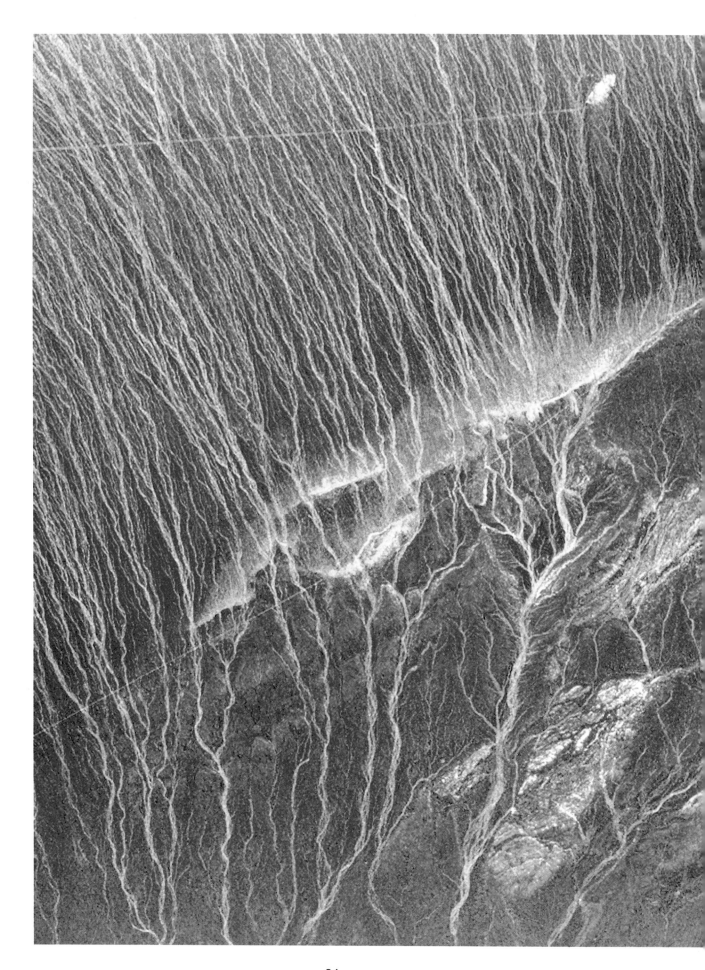

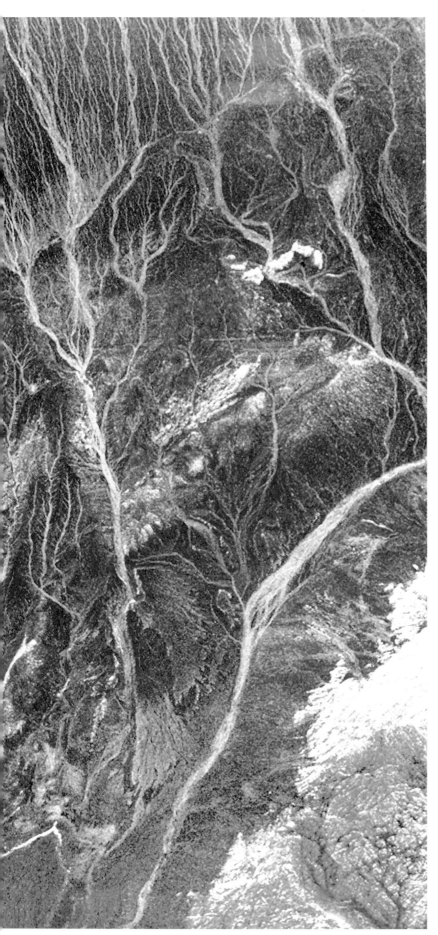

0   20km

SIR-C image of a groundwater enrichment belt in Ejina
county (R: L-HH, G: L-HV, B: C-HV)

Soil Moisture

Soil moisture detection is meaningful for agricultural, hydrological, and meteorological research and for their related operational purposes. For agriculture, soil moisture information is helpful for predicting crop yields and plant diseases. For hydrology, it is an important parameter for calculating precipitation seepage and irrigation management. For meteorology, it is a critical factor for calculating solar radiation energy.

Microwave is sensitive to soil dielectric properties and consequently to soil moisture. Since dielectric property is an important parameter for determining the target backscatter coefficient, there exists a direct relationship between the surface backscatter coefficient and soil moisture.

In the microwave band, the dielectric constant of water is 80. For dry soil, it is from 3 to 5, and for wet soil, it can reach more than 20. This shows the sensitivity of microwave to soil moisture.

An experimental relationship between soil moisture and backscatter coefficient was established through the calibrated airborne

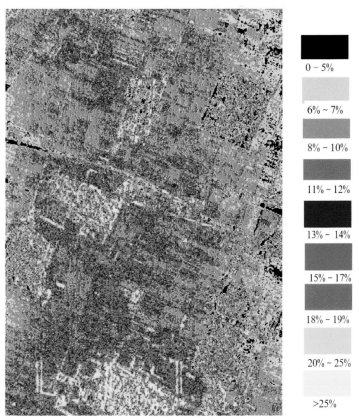

Water content derived from backscattering coefficients

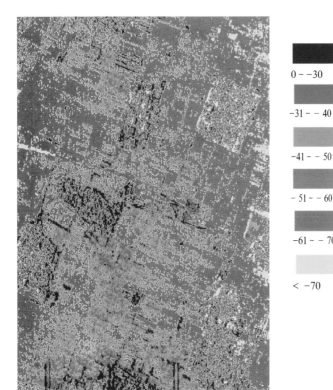

Backscattering coefficients image

SAR image of nearby Shijiazhuang. Thus soil moisture can be calculated from image brightness values or grey-levels.

The SIR-B image of the Huangyangzhen area, Ganshu Province acquired on November 17, 1986 shows a pie-like field with a bright tone in the center of the image. It is a recently irrigated field and yields a contrast to other fields. Since there is no obvious difference of surface roughness between these fields, the difference of soil moisture is a critical factor causing this contrast. In fact, by the presence of a dark slice of the pie-like field, it appears that a circular pivot irrigation system was operating during the acquisition of the radar data (Referred to "The Final Report of Spaceborne SAR Applications: Hydrological Application").

SIR-B image of Huangyangzhen area, Gansu province

GLACIER

Glaciers of Northeastern Aksayqin Lake

The valley glaciers of western Kunlun Mountains, to the northeast of Aksayqin Lake are particularly visible on this spectacular SIR-C multi-band, multi-polarization false-color composite image, and these glaciers show up as blue-purple color. At the front of glaciers 1,5 and 9, due to retreat, grows hummock-and-hollow moraine hills are forming which show up as orange color. At the front of glaciers 2 and 4, ice-barrier lakes enclosed by two end moraine dykes are forming. Since the lake ice is clear and its surface is smooth, radar wave can penetrate the ice layer to detect the end moraine underneath and yield a "fog-covered" orange color. The side moraines of glaciers 1, 2, 3 and 4 show up as a yellow color and are easily discriminated. There are still many cirque glaciers in addition to these valley glaciers in the up-left part of the image, showing up as a white color mottled with cyan speckles. This special image feature is caused by the radar backscatter mechanism: L-HH backscatter intensity is mainly determined by surface scatter. The difference of local relief causes the change of local incidence angle, which is a key factor for determining the intensity of surface scatter, so the flecky tone on L-HH image is caused by relief undulation. L-HV and C-HV backscatter intensities are so strong due to volume scattering from the recrystalized snow or firn, that pixels' gray level are saturated and cannot reflect local relief.

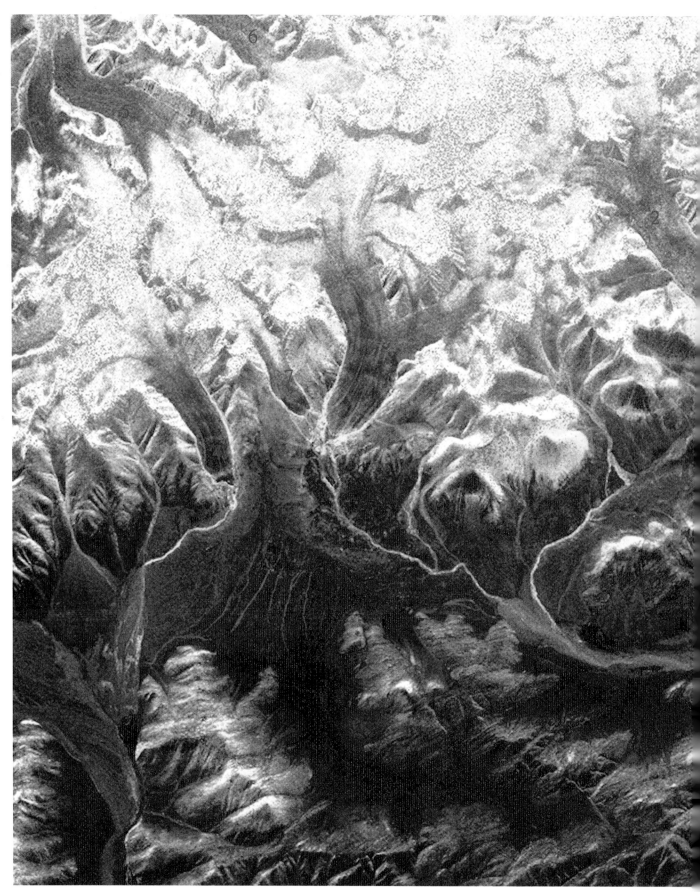

SIR-C Image Showing Glaciers to the Northeast of Aksayqin Lake

0   10km

Glaciers of Southeastern Qiaogeli Peak

The study area locates to the southeast of Qiaogeli peak, which stands on the border of China and Pakistan. This area belongs to the middle segment of the Kalakunlun Mountains, with an average elevation of 7,000 meters, and is characteristic of glaciers and glacier landform that are clearly visible on SIR-A image.

There are two main types of glaciers, valley glacier and cirque glacier, can be distinguished on SIR-A image. Cirque glacier is the small glacier formed in amphitheater shaped basin filled with firn, as is evident in J3-K3 and G3-H3. Valley glacier is the bigger glacier that can moves slowly along valley, as is evident in C1-C4, A1-A2 and H2-K2, among which C1-C4 is the biggest in this image. Controlled by geologic structure and topography, the flowing direction of most of these glaciers is NW-SE and NE-SW.

A variety of glacier landforms has been sculptured or deposited by glaciation, including cirques, glacier valleys, matterhorns, aretes and moraines. All of these landforms are easy to be discriminated on SIR-A image. A cirque or firn basin is chair-like basin formed through freeze-and-thaw erosion at the source of valley or cirque glacier. A glacier valley is the straight valley with U-shaped cross section eroded by a valley glacier. What is shown up in C1-C4 and I2-K2 are large-scale glacier valleys and what is shown up in F1-G1 and H1 belong to small-scale glacier valleys. An arete or knife-edge crest is the steep ridge between valley glaciers and cirque glaciers, as is evident in B4,B5,G5 and H4. A matterhorn or horn peak is the spiking pyramidal peak where several firn basins joint at their headwalls, as is easily discriminated in A1, B1 and H5. Moraines can be divided into lateral or side moraine, medial or mid moraine and terminal or end moraine according to their position in glacier. At A1/A2, A3-C4, E2-H4 long mid moraines are visible, ant at E2-H4, there are several mid moraines in one glacier. End moraines are formed at the terminal of a glacier, where ice thaws into water. For ice and water has completely different backscatter feature, end moraines are easily discriminated according to their position on radar image.

1. Firn basin
2. Knife-edge crest
3. Glacier valley
4. Glacier
5. Horn peak
6. Middle moraine
7. End moraine

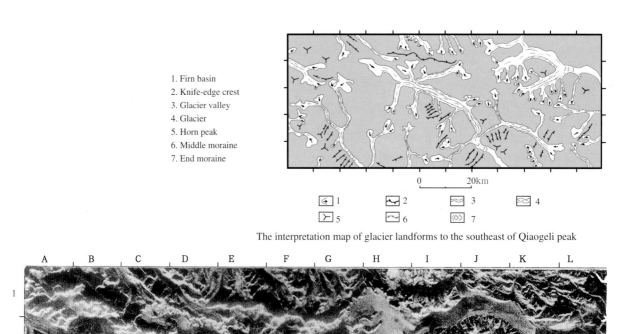

The interpretation map of glacier landforms to the southeast of Qiaogeli peak

SIR-A image to the southeast of Qiaogeli peak (**Illumination** from the bottom to the top of the image)

GEOLOGY

The most striking feature of a radar image is its accentuation of surface topography. The stereo effects of an imaging radar image from radar shadows combined with its sensitivity to the surface roughness enable it to be a very useful tool for geological mapping. In particular, it is a useful tool for frequently cloudy and rainy areas.

The earliest application of SAR images to geology can be dated back to the 1960's. Since then both airborne and spaceborne SAR have been widely applied to geological mapping and mineral exploration and have demonstrated the importance of their roles in geological applications. Especially, spaceborne SAR data has shown its advantage in geological studies owing to the rapidly imaging for a large area and less geometric distortion in comparison to airborne SAR data. With the development of radar remote sensing technology, a new discipline - Radar Geology - has been formed based on microwave scattering theory of geological surfaces as well as radar imaging, data processing, and geological analysis. This section consists of five parts, including typical terrain features, structures, lithological analysis, mineral exploration, and radar penetration studies. It covers 12 different types of airborne and spaceborne radar imagery, in which the most advanced imaging radar SIR-C/X-SAR images are dominant. In data processing, both qualitative and quantitative analyses have been applied to SAR data, including multifrequency, multipolarization SAR data as well as interferometric SAR data. The variety of radar images presented here demonstrates the capabilities of SAR for geological analysis.

TYPICAL TERRAIN FEATURES

Topography

Three Gorges Area

A very steep V-shaped valley is formed along both banks of the Yangtze River in the Three Gorges Area due to intensive uplifting of the crust. The surface has been incised up to 700 m. Over ten places along the Three Gorges section, the river bed is approximately 30m to 40m below sea level, indicating the extent of deep erosion of the river caused by crustal uplifting. On the digital elevation model (DEM) image produced by NASA/JPL using the interferometric SAR technique and Chinese airborne L-band SAR image, topographic features of the Bayan Gorge, one of the Three Gorges, are exhibited.

DEM image derived from interferometric SAR data (NASA/JPL).

L-SAR Image of the Three Gorges Area.

Karakax River Area

DEM information for the Karakax River area of the Western Kunlun Mountains was extracted from L-band VV polarization interferometric SAR (INSAR) data. The extraction of 3-D topographic information from INSAR data consisted of the following procedures: 1) processing of INSAR signal data, 2) normalization of imaging parameters, 3) precise data geometric registration, 4) phase correction for flat terrain, and 5) phase unwrapping. After data processing through the above procedures, a slant range DEM was obtained. Imaging parameters can be used to calculate the ortho-rectified DEM. It can be seen from the phase image and phase coherent image that the variation of phase is well matched with the DEM.

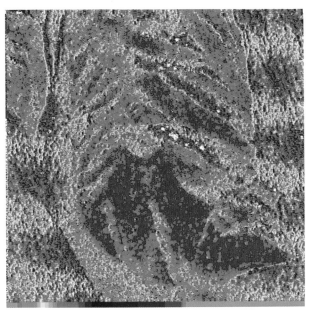

INSAR phase image
(The color bar represents variation from -180⁰ to 180⁰).

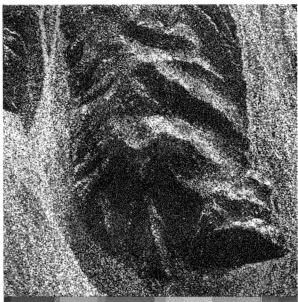

INSAR phase coherent image
(The color bar represents variation from 0 to 1).

DEM extracted from INSAR data.

3-D display of the DEM.

Typical Geomorphology

Loess

Loess is windblown dust of Pleistocene age, carried from desert surface, alluvial valleys, and outwash plains. The loess plateau of central China is probably the greatest single compilation of terrestrial eolian silts. The loess mantles about 320,000 km^2 in the provinces of Shanxi, Shaanxi, Gansu, and Ningxia. It thickens markedly to the east, where the thickness in the areas of Lanzhou and the north of Shaanxi Province reaches 100 to 300m. The loess plateau is being dissected by headwardly extending gullies. This feature is shown on a SIR-B image and an X-SAR image acquired on April 10, 1994. On the SIR-B image, the bright feature at the middle left is the city of Xining, the capital city of Qinghai Province. The areas shown on the X-SAR image are located in the vicinity of the boundary between Shanxi and Ningxia provinces.

Photo of loess landscape.

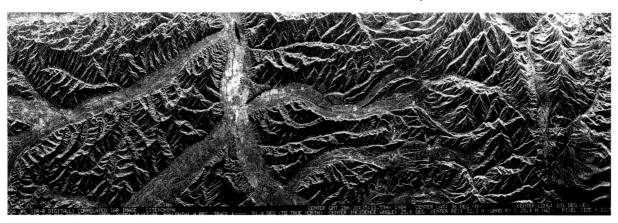

SIR-B image of loess geomorphology in the middle part of Qinghai Province (NASA/JPL).

X-SAR image of loess in northern Shaanxi Province.

0   10km

Karst Geomorphology

Solution topography is best developed in south China. This portion of south China encompasses part of the most spectacular karst topography found anywhere on the Earth. This scene in the southwest of Hunan Province demonstrates the capacity of SIR-C radar to accentuate karst morphology. In an oval basin at the center of the image, the exposed karst displays the peak-cluster haystack morphology and its relationship to regional structures. The development of these haystacks is obviously controlled by NE and NW linear structures. Other karst areas on the image are dominated by residual hills, which have been eroded by the drainage systems.

SIR-C false color composite image (R: L-HH, G: L-HV, B: C-HV) shows karst geomorphology in Daoxian County , Hunan Province.

0   2km

Sand Dunes

The Badain Jaran Desert in Inner Mongolia is the third largest desert in China and contains the world's largest sand dunes, up to 300 m high. Migration of barchanoidal sand dunes is a major characteristic of the desert. Migrating sand dunes comprise approximately 83% of the total area of the desert. Sand dunes are in linear chains of barchanoidal/transverse dunes and oriented in a N30°E to N40°E direction, reflecting the influence of NW wind to the area.

Radarsat image shows sand dune morphology of the Badain Jaran Desert.

0   50km

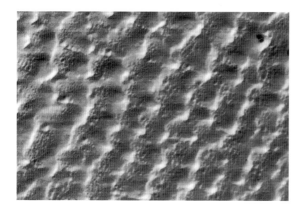

SIR-A image of the Badain Jaran Desert.

Landsat MSS image of the Badain Jaran Desert.

On the Radarsat ScanSAR image, the entire Badain Jaran Desert is shown. The morphology and orientation of the sand dunes are obvious. Since Radarsat imaged the area in an ascending orbit, the radar look direction is nearly perpendicular to the sand dune chains, which resulted in a bright radar return. In contrast, on the SIR-A image, very little information on dune structure is visible. This is because sand is smooth to L-band radar, and only dune slopes facing south return energy to an antenna. The main trend of dunes is parallel to the look direction; thus the whole area appears mainly dark.

Alluvial Fans

The Yumen Alluvial Fan

This Radarsat image scene was taken on December 23, 1996 and shows a large alluvial fan in the Hexi Corridor of Gansu Province, approximately 500 km east of Lop Nur. The fan is derived from flash flood-like water flowing out of the Qilian Mountains at the bottom of the image. Active drainage has entrenched an older, darker fan surface, which was elevated to 1200m to 1300m and is composed mostly of gravelly gobi. The formation of the fan is closely related to the uplift of the Qilian Mountains during the Quaternary and subsidence of the basin in front of the mountain range. In this area, neotectonic movement is very active and is manifested by the revival of old faults as well as newly formed scarps, together with frequently occurring earthquakes.

The front edge of the fan is an outflow area of groundwater; thus, soil moisture is high and results in high radar return. The Shule River originated from the Qilian Mountains flows along the eastern edge of the alluvial fan and then to the west, irrigating the farmlands in the north portion of the scene.

Radarsat image of the Yumen alluvial fan.

0   20km

Alluvial Fans in Karakax Valley

The Karakax Valley is at 3600 to 4000 m elevation and is largely controlled by the Altyn Tagh Fault. The valley's current climate is hyperarid with almost no vegetation, except along the river floodplain and along a few glacier-fed tributaries. Strong winds from the northeast have scoured the valley, producing numerous ventifacts and depositing silt. Rock types in the mountains bounding the valley consist mainly of metasedimentary and plutonic rocks.

Three alluvial fan units can be identified on the SIR-C image of the Kaugaiwar Valley area in the Western Kunlun Mountains. Farr and Chadwick (1996), in their detailed study of radar signatures of alluvial fans, compared the alluvial fans of Death Valley, California to those of the Kunlun Mountains in order to reconstruct the palaeoclimitic history of the Kunlun mountains. Geomorphic processes affecting alluvial fans in the two areas include aeolian deposition, desert varnish, and fluvial dissection. However, salt weathering is a much more important process in the Kunlun than in the southwestern United States. This slows the formation of desert varnish and prevents desert pavement from

forming. Thus the Kunlun signatures are indicative of the dominance of salt weathering, while signatures from the southwestern United States are characteristic of the dominance of desert varnish and pavement processes. Remote sensing signatures are consistent enough in these two regions to be used for mapping fan units over large areas. The oldest fan units are typically highest and dissected by the lower, younger deposits. In some cases, younger units are entrenched into the older and deposited at the distal end of the fan; however, in others, the younger units are deposited to the side of the older units, which have been moved laterally away from the active stream channel by strike-slip movement on the Altyn Tagh Fault.

The youngest fan unit is composed of fresh cobbles and boulders. The rocks on this unit are composed of coarse-grained siliceous granitic, fine-grained mafic and metamorphic rock types. The rocks are slightly weathered, with a slight staining of desert varnish on the less recently active surfaces. Rocks cover about 50% to 80% of the surface, with rock diameters ranging up to 35 cm. Since the surface is rough, it gives rise to strong radar echo and therefore gives a bright radar return. Situated about three meters above the most recent depositional unit, an intermediate fan unit can be mapped on the basis of its bouldery surface and the presence of desert vanishes on the more resistant boulders. This fan

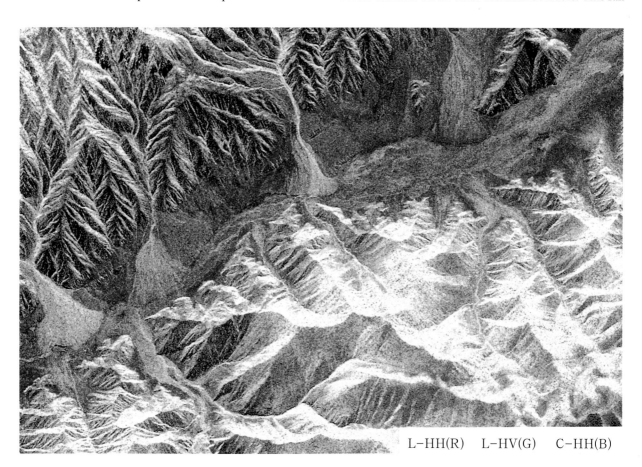

L−HH(R)   L−HV(G)   C−HH(B)

SIR-C false color composite image (R: L-HH, G: L-HV, B: C-HH), showing the Karakax Fault and alluvial fans.　　0　　　　　　　3km

TM7(R)   TM5(G)   TM1(B)

Landsat TM false color composite image of Karakax Fault and alluvial fans.

Field photographs taken from alluvial fan sites.

unit is a relatively dark color on the SIR-C image. The oldest unit on the alluvial fan is generally situated six meters above the intermediate unit and is recognized principally by the lack of exposed rocks. Rock relief is only about one cm, and the rock cover is about 30%, composed almost entirely of pebbles and flakes of fine-grained rocks with light to moderate varnish. This unit is darkest on the SIR-C image. A comparison is made with the SIR-C image and the Landsat TM image. It shows that the SIR-C image is better for discriminating the alluvial fan units since radar is sensitive to surface roughness.

Pishan Alluvial Fan

At the foot of the Kunlun Mountains in Xinjiang, alluvial fans of different units are developed. They show evidence of paleoclimatic changes in geological history. This false color SIR-C image is a color composite of L-HH (R), L-HV (G) and C-HV (B). The image center is

SIR-C false color image of the Pishan alluvial fans at the foot of the Kunlun Mountains, Xinjing (R: L-HH, G: L-HV, B: C-HH).

located at 37°18.3′ N, 78°28.2′ E (latitude/longitude). Incident angle at the image center is 52.3°. The fernleaf like alluvial fan running horizontally across the middle of the image is the oldest unit of alluvial deposition, indicating that heavy rainfall or glacial meltwater resulting from rising global temperatures deposited a large amount of coarse materials at the foot of the mountain. Although the fans

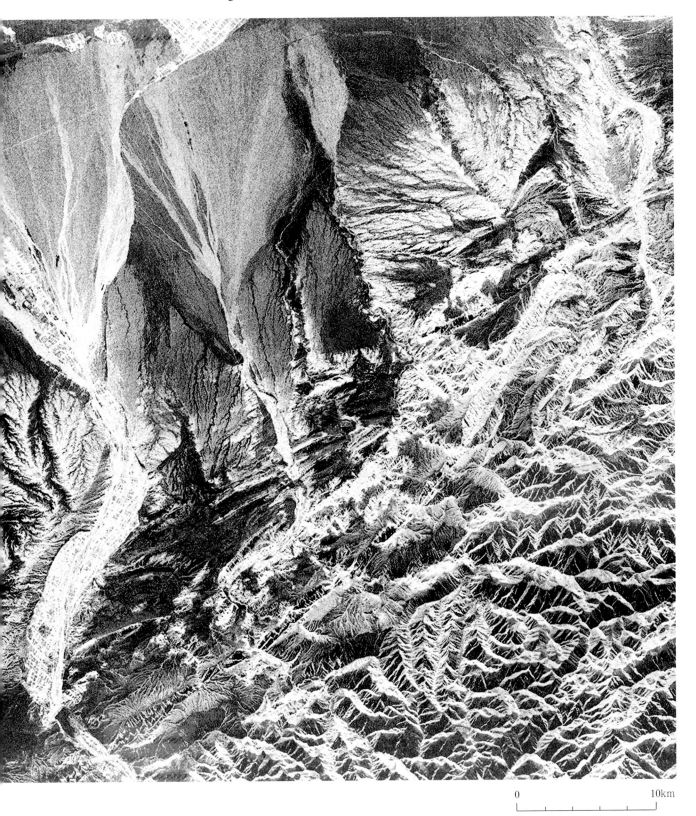

0   10km

were deposited in an early time, they give relatively strong radar return due to the complex morphology formed by the dissection of later streams. The intermediate surfaces of the alluvial fans are relatively dark on the radar image since the flat surface consists of fine weathered materials. A young or alluvial fan surface is light purple in color, indicating a rough surface. It can be seen from the morphology of alluvial fans that the formation of the fans was controlled by tectonic movement, with uplifting occurring in the SE and subsidence in the NW. The most recent sedimentary deposits are distributed along the river valleys, finally forming an oasis at the outreach of the fans. The bright lattice features are produced by white poplars, which form wind shelters surrounding irrigated farmlands.

The ERS-1 SAR image shows the left part of the SIR-C image from which different alluvial fan units can be identified.

Black and white ERS-1 SAR image of the Pishan alluvial fans (provided by Dr. T. G. Farr of NASA/JPL).

Volcanoes

Kunlun Mountain Volcanoes

A group of volcanoes northeast of Aksayqin Lake in the western Kunlun Mountains have been identified on SIR-C/X-SAR imagery acquired on April 17, 1994. The volcanoes are at an elevation over 5300m above sea level. There are well-exposed volcanic cones and lava flows in the area. Scientists from the Institute of Remote Sensing Applications, Chinese Academy of Sciences and NASA/JPL jointly conducted field studies in this area in August 1995. Two types of lava flow, pahoehoe and blocky lavas, and nine cinder cones were mapped, which had not previously been reported. On the geological interpretation map, cinder cones 1, 2, 3 and 4 are easily identified because they occur in an open valley and have high topographic relief; the other five cones have positive topography and circular form, but they occur on top of the lava flows and therefore are not easily detected. The squint SAR geometry accentuates the cinder cones and enables them to be identified. Most lava flow areas are occupied by pahoehoe lava, which forms a relatively smooth surface. The surface of pahoehoe flow is rather flat and mainly composed of smaller fragments about 10 to 20 cm in size. The radar signature of this type of flow appears relatively dark due to specular scattering from the surface, whereas radar return from the toe of the flow appears very bright, particularly when the toe faces toward the radar illumination. The blocky lava flows are distributed mainly around cinder cones 5, 6, 7 and 8. Their surface is noticeably rough with irregular fragments of clinker ranging from 30 cm to more than 1m size. The lavas have been eroded and weathered. Vesicular and amygdaloidal rocks are the most obvious features on the flows. The rough surfaces give rise to strong diffuse scattering of the electromagnetic wave.

Comparison of five black and white wavelength and polarization combinations of SIR-C/X-SAR images shows that L- band HV polarization is the best one for identifying cinder cones and for discriminating the two different types of lava flows, and discriminating lava flows from alluvium and bedrock. This conclusion can also be drawn from the diagram and table for comparing σ^0 values of pahoehoe and blocky lava, alluvium, and bedrock at different wavelengths and polarizations. The backscatter coefficients (σ^0) were extracted for a number of homogeneous areas of pahoehoe and aa lavas, alluvium and bedrocks. The L-HH image has good ability to differentiate aa lava from the other three types; however, the contrast between pahoehoe, alluvium, and bedrock is not very large. Similarly, the C-HH image has a relative good ability to distinguish aa lava from the other surfaces, as well as bedrock from alluvium and pahoehoe lava; however, pahoehoe lava and alluvium

3-D display of volcanoes northeast of Aksayqin Lake.

False color SIR-C image of the volcanoes northeast of Aksayqin Lake (Left, R: L-HH, G: L-HV, B: C-HV; Right, R: L-HH, G: L-HV, B: C-HV).

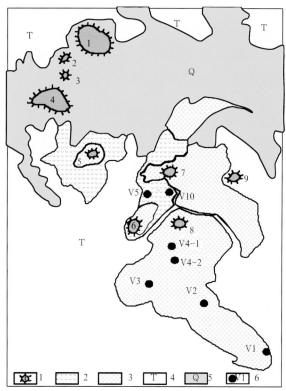

Geological Interpretation Map of SIR-C images:
1. Cinder cones, 2. A'a lava, 3. Pahoehoe lava.
4. Triassic strata, 5. Quaternary, 6. Sample sites.

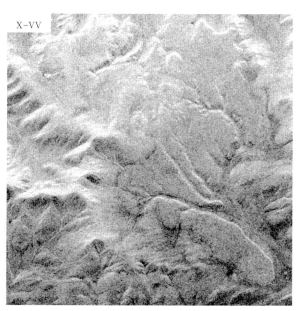

Comparison of five black and white wavelength and polarization combinations (this page and the next) of SIR-C images indicating L-HV has the best effect for discriminating different terrain features.

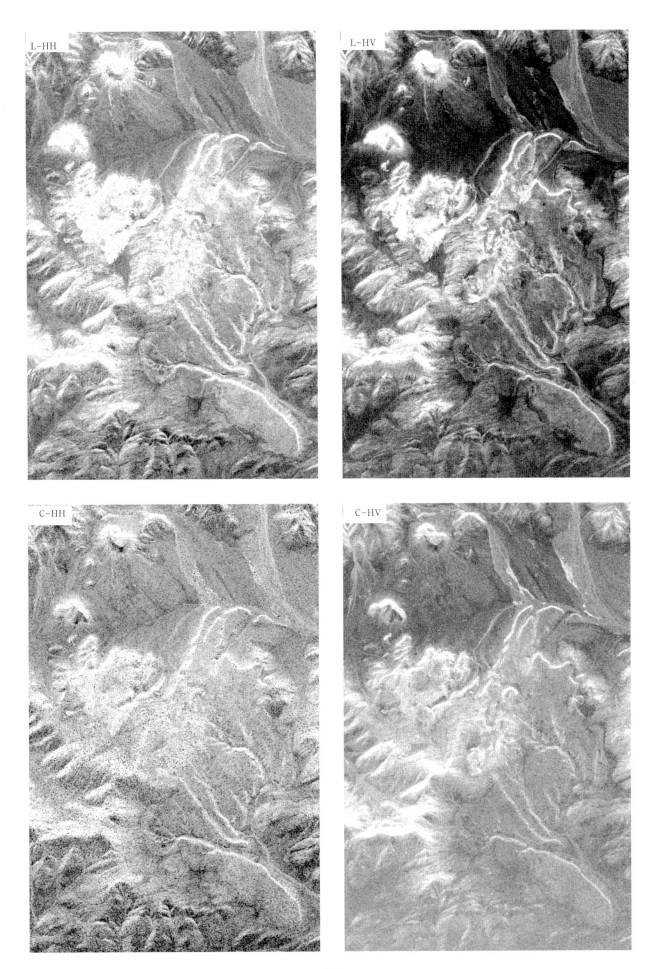

Photo of the largest volcanic cone.

Photo of pahoehoe surface.

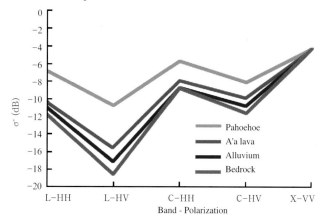

Comparison of backscatter coefficients of different terrain

In this study, eight volcanic samples were collected in the field (from sample locations, see geological interpretation map), and a coaxial probe in a microwave network analyzer measured their complex dielectric constants. Because the measured imaginary part is very small, only real part is plotted in the figure. The range of dielectric constant for these rocks is 3 to 8, which corresponds to measurements of basaltic rocks by Ulaby et al. (1988). It can be seen from the figure that there are two groups of dielectric constant: the upper group represents denser basaltic rocks; the lower group is associated with vesicular rocks. The results suggest that dielectric constant is relatively independent of frequency and mainly relates to the density of rocks.

Geochemical analyses of volcanic rock samples collected from the field indicate that the potassium content is very high. K_2O is commonly higher than 4%, and $K_2O > Na_2O$. The samples are enriched in the trace elements Rb, Sr, Ba, Nb, and Ta and in light rare earth elements. They basically do not show Eu anomalies. The rocks can be classified into the shoshonitic series, which is associated with intracontinental subduction. The K-Ar isotopic ages range from 7.45 to 3.97 Ma, indicating there were at least two volcanic eruptive phases. This gives further evidence for the presence of neotectonism in the Qinghai-Tibetan plateau since the Cenozoic era.

Contact of two lava flow types.

Photo of blocky lava.

are not distinguishable, because the values for these two types of features fall at the same point. Similar to the L-HH image, the C-HV image has a good ability to separate aa lava from the other three surface units, and there is also a slightly difference between pahoehoe lava, alluvium, and bedrock. In contrast with the L- and C-band images, the surface units are nearly identical on the X-VV image, whereas the σ^0 values show little variation.

Digital Number (DN) and σ^0 Value from SIR-C/X-SAR Data for Lava Flows, Alluvium, and Bedrock

Lava type (pixels)	Band and Polarization	DN Range	DN Mean	DN SD	σ^0(dB)
Pahoehoe Lava (10,378)	L-HH	0-215	78.081	25.495	-11.07
	L-HV	12-117	40.801	13.494	-17.24
	C-HH	0-219	71.007	29.786	-8.71
	C-HV	19-148	64.998	17.234	-10.85
	X-VV	1-22	7.866	2.655	-4.29
Aa Lava (7,296)	L-HH	0-255	206.234	55.982	-6.85
	L-HV	32-255	179.230	64.397	-10.81
	C-HH	0-255	140.589	57.949	-5.74
	C-HV	31-255	122.229	42.369	-8.10
	X-VV	4-40	13.639	4.296	-4.35
Alluvium (6,270)	L-HH	13-255	67.645	23.478	-11.70
	L-HV	6-134	30.004	10.615	-18.58
	C-HH	0-245	74.073	27.051	-8.52
	C-HV	11-159	55.980	17.132	-11.50
	X-VV	3-26	10.296	3.093	-4.31
Bedrock (10,913)	L-HH	0-255	91.239	51.943	-10.40
	L-HV	8-255	59.998	42.176	-15.57
	C-HH	0-255	85.105	46.499	-7.92
	C-HV	20-255	81.782	31.552	-9.85
	X-VV	2-33	9.779	4.291	-4.31

Lithology and K-Ar Isotopic Ages of Volcanic Samples

Sample no.	Lithology	Isotope age (Ma)
V1	Iddingsitive basalt	7.45 ± 0.19
V2	Vesicular basalt	5.40 ± 0.16
V3	Vesicular pyroxene basalt	4.70 ± 0.14
V4-1	Basalt	3.97 ± 0.12
V4-2	Pumice	6.77 ± 0.17
V5	Vesicular basalt	5.78 ± 0.15
V10	Trachyte	5.11 ± 0.15

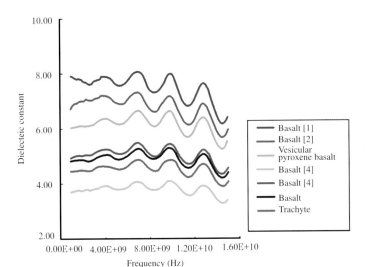

Real part of the dielectric constant for volcanic rocks at different frequencies. The upper group represents denser, nonvesicular rocks, and the lower group represents vesicular rocks.

Hainan and Leizhou Peninsula Volcanoes

Hainan Island and the Leizhou Peninsula in south China have many similarities in respect to their geological settings. They are separated by the Qiongzhou Strait. The areas are mainly controlled by NW, NNE and near EW trending structures. A prominent feature in both areas is the existence of Quaternary volcanoes, which erupted along NW and NEE directions.

For the Hainan volcanoes, the following geological information can be derived from the SIR-C/X-SAR images acquired in April 1994.

(1) Quaternary volcanoes erupted in many phases from Q_1 to Q_4. The volcanoes are described below.

βQ_2^1- βQ_2^2 Basalt erupted in the Miocene. It has been weathered into a thick red clay plateau. Most of the plateau surface has been used as dry farming land or grassland (photo 1). The tone on the SIR-C image is light. The fluvial-dissected valley is moist farmland; the tone is dark on the image.

υQ_2^1 Tuff erupted in the Miocene, which is now farmland.

βQ_4 Basalt plateau formed in the Holocene. The rocks are slightly weathered, and bare rocks are exposed. The surface is rough. A small amount of wind-blown materials fills in the gaps of the rocks (photo 2).

βQ_3^1- βQ_3^2 Basalt erupted in the Pliocene. The weathering situation is between βQ_2^2 and βQ_4. The surface is rough. The higher elevated basalt exposes bare rocks whereas the lower part is thick weathered sediment that is presently farmland (photo 3).

υQ_4 Tuff and pyroclastic rocks erupted in the Holocene. They are now farmland.

There are two types of volcanic vents. One exhibits a circular negative topography; it formed during a volcanic explosion that had a large amount of lava overflow. Another exhibits a positive cinder cones (photo 4).

On the SIR-C image, two types of plutonic bodies can be discerned. One is τ_5^B, i.e., biotite-granite, Its topography is dominated by hills. The weathered material is quite thick, and composed mainly of sandy soils. The soil is favorable for preserving water. Most of the relative flat area has been reclaimed to farmland. The land in the valley is used for paddy field, and the flat area on the slope is used for terrace land. Another type of plutonic body is γ_5^C. It has a circular feature on the image, which maybe an implication of the difference between a new type of plutonic rock body of γ_5^C and country rocks, or represents lithological phase variation within γ_5^C.

The metamorphic rocks are mainly composed of Devonian and Carboniferous (D-C) slate and phyllite (photo 5). These rocks form hilly terrain. On the surface is weathered red clay material. The soil has poor water-holding capacity, therefore, it is very dry, and not

SIR-C image of the Chengmai area, Hainan Province (R: L-HH, G: L-HV, B: C-HH). Below are airphotos of different features.

Photo 5

Photo 3

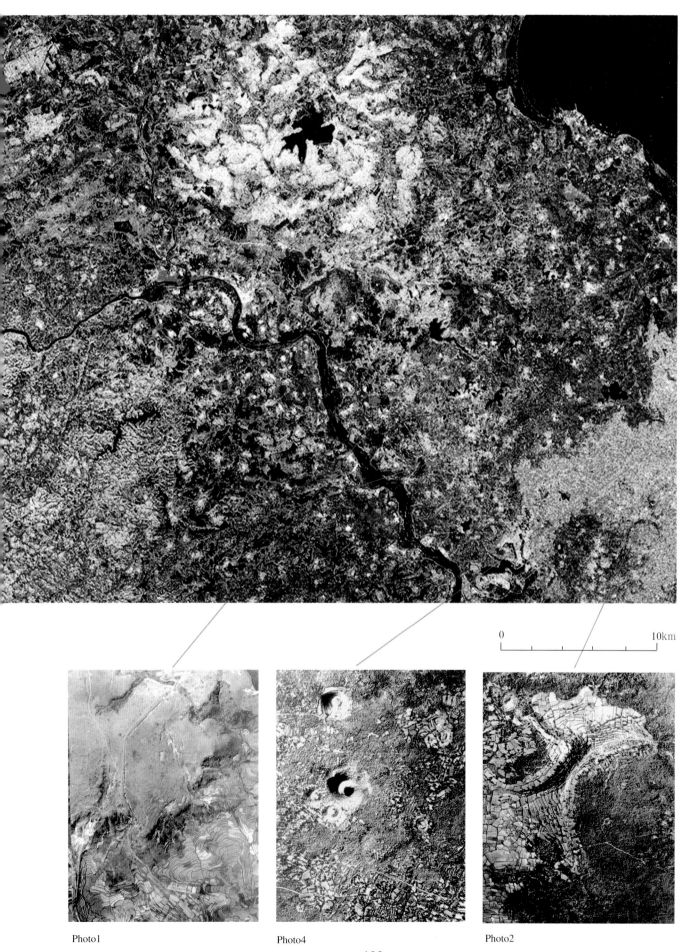

0   10km

Photo1   Photo4   Photo2

favorable for farming.

The sedimentary rock is mainly composed of Cretaceous sandy gravel (K_2). It has a relative loose texure. Most of the area is bench terrace, and most of them have been reclaimed to be farmland.

The comparison of linear stretched L-HH, L-HV, C-HH, C-HV and X-SAR images suggests that L-HV can best discriminate volcanic features. This suggests that depolarization plays an important role in geologic analysis.On the X-SAR image of Hainan area, it is difficult to see the difference of lava flow, but the vocanic vent features in positive and negative topography can be delineated.

On the X-SAR image of the Leizhou Peninsula, volcanic cones and lava flows can be identified. For example, note the volcanic cones at E4 and G7 and the lava flows around F4 and G4. These features suggest that the Leizou Peninsula is an area of frequent volcanic activity.

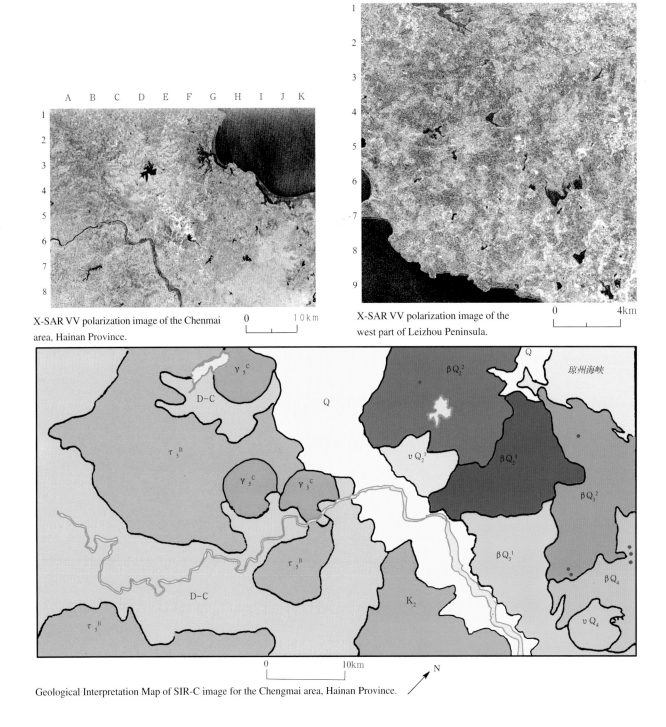

X-SAR VV polarization image of the Chenmai area, Hainan Province.

X-SAR VV polarization image of the west part of Leizhou Peninsula.

Geological Interpretation Map of SIR-C image for the Chengmai area, Hainan Province.

110

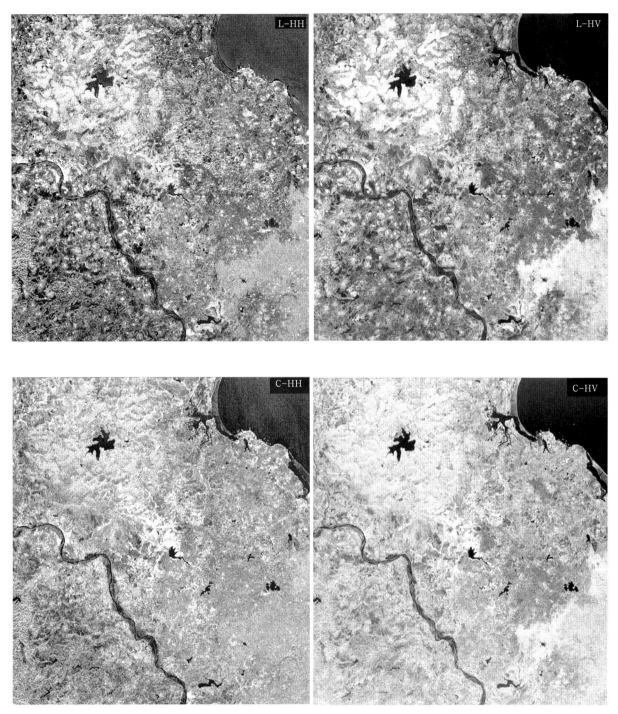

Comparison of different wavelength and polarization combinations for SIR-C images of the
Chengmai area, Hainan Province.

0   10km

Datun Volcanoes, in the North of Taiwan

The Datun Volcanoes occur in the utmost northern part of Taiwan Island and about 15 km to the south of Taibei. There are 20 volcanoes; among them is the Qixingshan Volcano, the highest and youngest volcano. Phenomena such as hot springs, blowholes, and solfataras are often seen. Most of the volcanic craters are well preserved. These volcano groups have erupted continuously, alternately spewing andesite, volcanic ash, and pryroclastics. The volcanic sediments overlie sedimentary layers of the Miocene at different ages. Most of the andesitic lava is comprised of pyroxene andesite, hornblende andesite, hyperite andesite, or a mixture of these three types of rocks.

On the false color SIR-C image composed of L-HH (R), L-HV (G) and C-HV (B), the Datun volcanic area is in purple. Volcanic cones can be identified from the image.

0   5km

False color SIR-C image showing Datun Volcanoes of northern Taiwan (R: L-HH, G: L-HV, B: C-HV).

Changbai Mountain Volcanoes

The "Heaven Lake" volcano, at an elevation of 2300 m and is situated at the peak of Changbai Mountains on the border of China and Korea. This is a famous large composite volcano. Surrounding the volcano is a lava plateau. Subsequent to the collapse of the top of the volcanic cone, which formed a large caldera, a spectacular crater lake called the Heaven Lake was formed.

Volcanoes erupted several times from the Neogene to the Quaternary. Early eruptions were dominated by fissure eruptions. Later volcanic activities were mainly central eruptions. The intensity of volcanic eruptions gradually weakened, and the amount of erupted lava was reduced. The lava changed from basic to alkali, resulting in an increase of viscosity and reduction the flow covered area. Towards the end, a large composite volcano was formed on top of the lava plateau. According to historic records, this volcano erupted in 1597, 1668 and 1702.

On the Landsat TM false color image acquired on May 14, 1985, generated from TM bands 5, 4 and 3 (R, G, B), the blue circular feature is called "Heaven Lake." Radiating unweathered basalt surrounds the lake. The red-brown area is covered by weathered materials and pyroclastics. Grass is yellow, and forests are green. From the circular features and tone of the image, it can be

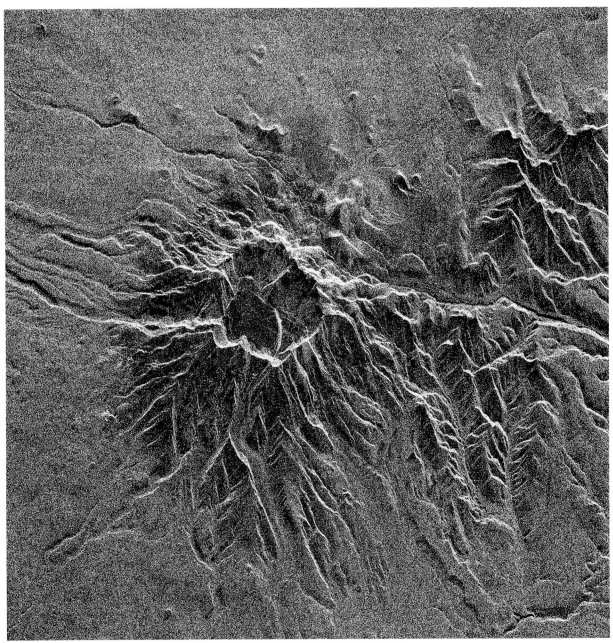

ERS-1 SAR image showing Mt. Changbai Volcanoes.

Illumination

N

0   3km

Landsat TM image of Mt. Changbai Volcanoes.

inferred that the lower base of "Heaven Lake" volcano was formed by multiple eruptions. From tonal differences between the east and other areas, it can be concluded that the drift direction of volcanic ash of the most recent eruption is towards the southeast.

On the black and white ERS-1 SAR image dated December 24, 1995, the enhanced volcanic features are very obvious due the side-looking geometry layover and shadowing of the SAR image.

STRUCTURE

Faults

Kangxiwar Fault

The Kangxiwar Fault (also called the Karakax Fault) is in the western part of the Altyn Tagh Fault, which is the most prominent fault of Tibet and western China, extending more than 1500 km from the Karakorum to the Nanshan. This is a gigantic left-lateral strike-slip fault, and also forms a boundary between the fold systems of western Kunlun and Karakorum. The fault, on both of its sides, has played an obvious role in controlling sedimentary construction, magmatic activity, metamorphism, and mineral deposit distribution.

On the false color composite SIR-C image acquired in April 1994, considerable evidence for recent movements along the fault is apparent from looking at details visible in the image. The most striking feature on the images displayed is the morphology of the triangular facet. Three different elevation levels of triangular facets can be discerned, each representing the rapid motion of the fault.

Space Shuttle Endeavor hand-held photograph showing the Kangxiwar Fault, which follows the valley and extends across the photograph to the right (provided by NASA/JSC/JPL).

3-D image generated from Interferometric SAR for a section of eastern Kangxiwar Fault (NASA/JPL).

X-SAR image acquired on Oct.7, 1994, showing the intersection of the Kangxiwar Fault with the Dahongliutan Fault.

0   10km

False color SIR-C composite image showing the Kangxiwar Fault (R: L-HH, G: L-HV, B: C-HV).

0   10km

Kalaxianger Fault

The Kalaxianger fault is situated in Fuyun and Qinghe Counties of the Altay Area, Xinjiang . The main landforms of the area are high relief terrain to the north and basin topography to the south. On the standard mode Radarsat image, the Kalaxianger Fault is oriented in a NNW direction and is clearly visible in the middle of the scene. Both walls of the northern section of the fault appear with concordant differential uplift, controlling the development of the Quaternary basin. In contrast, the walls of the southern section of the fault are in a reverse movement. The east wall was uplifted to form moderate relief mountains, and the west wall borders a Quaternary sedimentary basin.

This area is a conjuncture point for five major faults in the region. Therefore, this area is rich in minerals. The famous Kalatongke Cu-Ni deposit is located at the northern margin of a circular feature to the south of Fuyun County.

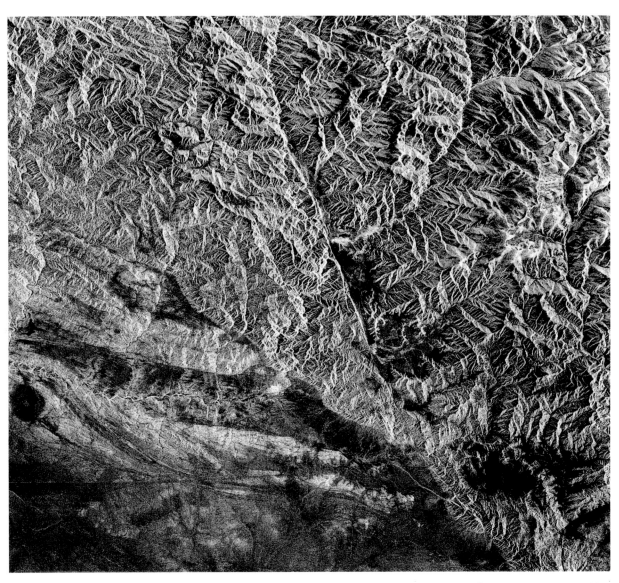

Radarsat image showing the NNW Kalaxianger Fault of Altai, Xinjiang .

0   10km

Faults in the Taiwan Region

Taiwan Island, the largest island in China, is situated in the central section of the island arc in the western Pacific Ocean. The Yushan Mountain Range in central Taiwan extends from the Sandiao Cape in the north to the Eluan Promontory in the south for a distance of about 350 km parallel with the Xueshan Mountain Range. To the west of the Yushan and Xueshan Mountain Ranges, a long and narrow rift called the Taidong Rift has developed, which is about 140 km from north to south and varies from 3 km to 6 km in width. The Taidong Rift is a suture of the collision between the Eurasian Plate to its west and the Philippine Sea Plate to its east; this is why frequent earthquakes have occurred along this rift.

Taiwan Island is a Tertiary geosyncline developed on top of a late-Palaeozoic to Mesozoic metamorphic rock basement. The mountain building movement has been very active since the Tertiary. The NNE oriented faults dominate the faulting system of the island and can be seen from the displayed airborne SAR image. On the false color SIR-C survey image that covers the center of the island, the geomorphology of the Taidong Rift is shown. The ALMAZ SAR image of the former USSR illustrates the faults in the Yilan area of northern Taiwan Island.

Black and white airborne SAR image of Taiwan Island (Provided by Energy and Resources Institute, Taiwan Academy of Industries).

Natural color ground photo of the landscape of the Taidong Rift.

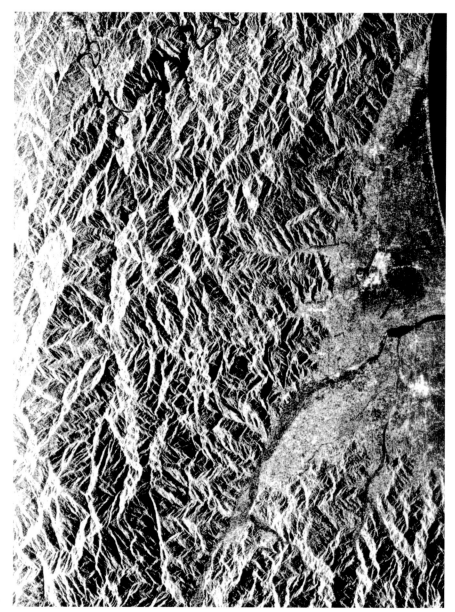

ALMAZ SAR of the Yilan area.

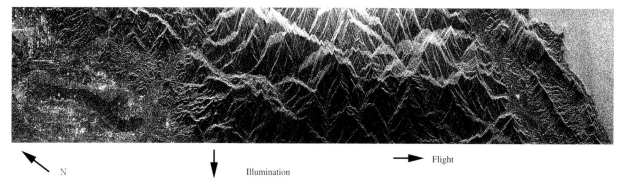

N

Illumination

Flight

SIR-C Survey Image (Mode 16) of Central Taiwan, showing characteristics of the Taidong Rift (right side of the image).

Dunhuang-Anxi Fault

This false color SIR-C image at coordinates 40°21′ 59″ N Latitude and 95°29′ 30″ E Longitude shows part of a fault extending from Dunhuang to Anxi.

The image was acquired in April 1994 and utilized an imaging Mode 11. Incident angle at the center is 52°17′ 02″ .

On the northwest wall of the NE Dunhuang to Anxi Fault, several alluvial fans of multiple phases were developed. The fans of mid-Pleistocene are clearly seen

False color SIR-C image (R: L-HH, G: L-HV, B: C-HV) showing a segment of the Dunhuang to Anxi Fault.

0   20km

and appear to have maintained good shapes. However, the fans are developed asymmetrically, and their middle axis migrated towards the left indicating an obvious deformation of the fans. The alluvial fans of the late-Pleistocene have also undergone deformation. From alluvial fan shapes, it can be seen that the nature of the fault is a left-lateral movement.

Altun Fault

The Altun Fault is the biggest fault in China and controls the northern boundary of the Qinghai-Tibet Plateau. The fault is composed of three minor faults, accompanied by a number of parallel branch faults.

On the remote sensing imagery, the faults are clearly shown, and linear structures are usually very obvious.

The scene of the SIR-C Mode 16 image, acquired in April 1994, shows a segment of the Altun Fault on the Qarqan River with coordinates at image center of 37°33′ 40″ N Latitude and 85°43′ 32″ E Longitude. There are considerable differences in the relief between the uplifting mountains and valley development to the east and west of the turning point of the L-shaped river. It can be seen from the image that a left-lateral strike-slip motion characterizes the fault.

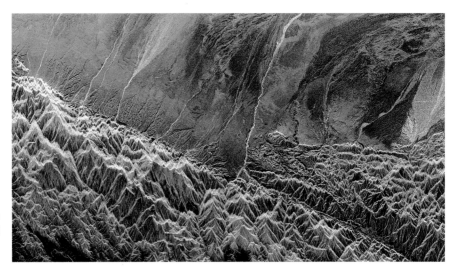

0   10km

False color SIR-C image showing a segment of the Altun Fault in Xinjiang (R: L-HH, G: L-HV, B: C-HV).

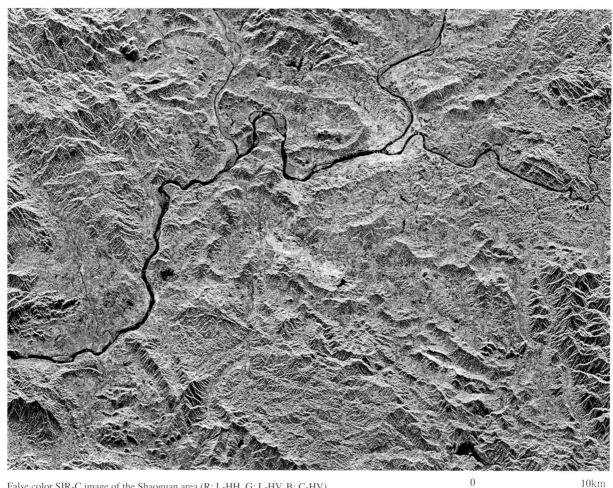

False color SIR-C image of the Shaoguan area (R: L-HH, G: L-HV, B: C-HV).

0   10km

The regional structure map of the Shaoguan area, Guangdong Province, indicates the existence of an EW structural belt. However, on the geological map of this area, the faults are dominated by NE-NNE structures. The false color SIR-C image is a composite of L-HH (R), L-HV (G) and C-HV (B), with coordinates at the image center of 113°41′ 32″ E Longitude and 24°48′ 38″ N Latitude. With high spatial resolution and side-looking characteristics, the false color SIR-C image reveals several parallel linear features and a giant acidic to intermediate intrusive rock belt of the Yanshan movement (J_2-K_2).

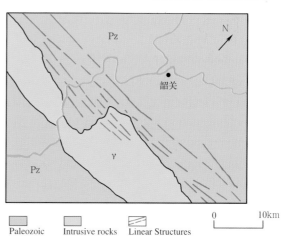

Paleozoic   Intrusive rocks   Linear Structures

0   10km

Geological Interpretation Map of the SIR-C image, Shaoguan area.

Ledong Fault, Hainan Province

This SIR-C/X-SAR image scene shows the fault related river valley geomorphology of Ledong to the Daiyun area, Hainan Province. Stratigraphically, the exposed rocks are considerably different on both sides of the river. To the north, the strata is dominated by Cretaceous rocks, while Devonian and Carboniferous rocks are confined in some local areas. To the south of the river, there are extensive areas of intrusive rocks of different ages. Due to differences in the components of intrusive rocks, these rock bodies exhibit different characteristics and texture on SIR-C image.

SIR-C image of the Ledong Fault, Hainan Province (R: L-HH, G: L-HV, B: C-HV).

0   20km

Structure of Mt. Langshan Area

This black and white SIR-B image of Inner Mongolia covers a portion of Mt. Langshan and two deserts: the Yamaleike Desert at the lower left and the UlanBuh Desert at the upper right. The rocks of Mt. Langshan are composed of Palaeozoic magmatitic granite and Precambrian gneiss and magmatite. The mountain range is in a "Z" shape and has a bright radar return due to the abundance of fractures and gullies on the rock surfaces that are oriented mostly in a NE-SW and NW-SE direction. There is a circular structure of negative topography on the uppermost left of the scene. On top of this feature, there is a very bright N-S linear structure. The bright return is from the corner reflector effect, in this case indicating the presence of a fault scarp caused by neotectonic movement. Other faults can also be easily mapped, even

when the faults pass through the deserts. Some sub-surface features of the rocks covered by sands have been revealed as a result of SAR penetration. These features are expressed in relatively bright radar return and have a regular shape on the black and white SIR-B imagery. On the Landsat MSS image of the same area, two sand belts (I, II), which have infiltrated into the UlanBuh Desert from the Yamaleike Desert, still have their imprint on the SIR-B image, clearly showing the distribution of rocks and fractures. Notable are the narrow dark strips on the SIR-B image which represent gullies ranging from

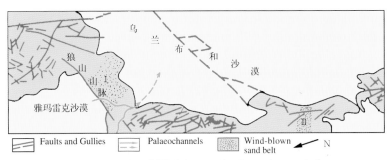

| Faults and Gullies | Palaeochannels | Wind-blown sand belt | N |

Geological Interpretation Map of SIR-B imagery for the southern part of Mt. Langshan, Inner Mongolia.

Black and white SIR-B image of the southern part of Mt. Langshan, Inner Mongolia.

125

several tens to one or two hundreds meters depth. Our preliminary studies suggest the existence of palaeochannels, as shown on the geological interpretation map of the SIR-B image.

Compresso-crushed belt, Qinghe

In the Qinghe area of the Altay district, Northern Xinjiang, the terrain is undulating and the climate is arid. On the airborne Earth Resource Radar (ERR) SAR image, the characteristics of tectonic fracture belts are clearly seen. The fracture belts are composed mainly of a group of large NWW linear structures representing regional tectonic framework characteristics. The large NWW linear structures are distributed in bright lines corresponding to linear mountain ridges and dark lines corresponding to mountain valleys on the ERR image. The fracture belts have the following characteristics:

1) Tectonic fracture belts are developed in the contact zone of different lithological units. Under the regional stress field, the relative movement of these rigid lithological blocks controls the occurrence and development of the compresso-crushed belt.

2) Tectonic fracture belts range from several meters to several hundreds meter in width. The belts are composed of radar bright and radar dark strips, and there are some irregular blocks in between the strips. These features reflect that the tectonic belts are composed of many different kinds of tectonic component.

3) The major direction of strike for fracture belts is NWW; however, in some places they appear in arcs. Some linear structures oriented in NE, NW and approximately NS intersect with or compound with the major fracture belts.

Black and white ERR SAR image showing the compresso-crushed belt of the Qinghe area, northern Xinjiang.

Folds

Fold in Western Kunlun Mountain Pediment

One of the manifestations of the collision between the Indian Plate and the Eurasian Plate is crustal shortening expressed by folding and thrusting. The neotectonic movement of the western Kunlun Mountains has apparent past intermittent activities. The intensity of the movement differs greatly from east to west and from north to south. In the east and west ends of the Kunlun Mountains, the uplifting of Pamir plateau in the west is more intensive than the uplifting in the east (> 1400 m). The differential movement in the mountainous area is manifested by the posthumous uplift and subsidence of the blocks. Whereas, a lateral compression is dominated in the piedmont of the Kunlun Mountains, resulting in neotectonic movements such as folding and thrusting. There are one to four rows of parallel folds at the foot of the western Kunlun Mountains. Their long axes are oriented generally in a 240° to 260° angle, and they intersect with the mountain range at angles of about 10° to 20°. These observations provide evidence for a regional right-lateral stress field reflecting that tectonic forces are moving the Kunlun Mountains to the north and are moving the Tarim Basin to the south.

On the false color SIR-C image, an anticline located southwest of Hotian at the foot of the West Kunlun Mountains is clearly seen. The core of the anticline is a Carboniferous stratum. Permian and Cretaceous strata are on both flanks of the anticline. However, the Permian and Cretaceous rocks have unconformable relationships. The core has strong radar return that differs from other strata. The asymmetrical characteristics of the anticline suggest that the forces acting on the anticline are unbalanced.

False color SIR-C image showing folds at the foot of the West Kunlun Mountains
(R: L-HH, G: L-HV, B: C-HV).

0   10km

False color SIR-C image showing Huoyanshan Anticline of Xinjiang (R: L-HH, G: L-HV, B: C-HV).

Huoyanshan Fold

The sinuous hinge anticline of Huoyanshan is situated in the east of Turpan County of the Xinjiang Autonomous Region. The core is composed of Jurassic and Cretaceous rocks, and the flanks are composed of Tertiary rocks. The axis of the anticline is a sinuous shape due to the effect of later NW and NE faults. The average strike of the fold is 295°. The anticline is pitched at two ends. The strata on the northern flank are well exposed. However, the outcrop strata are poorly exposed in the southern flank, most being covered by Quaternary sediments. Perhaps this is related to a strike-slip thrust fault near the core.

This false color SIR-C image scene has shown conspicuous features of bedded lithological units characterized by different color strips related to the variations of surface roughness of different rock types. The rocks are mainly of sandstone, conglomerate, and mudstone. The scene also illustrates alluvial fan materials derived from Mt. Tian Shan cutting through the fold and deposited at the south side of the anticline. On the single band and single polarization SAR images (L-HH, L-HV, C-HH and C-HV), the sensitivity of different bands and polarizations to different lithological units is clearly seen.

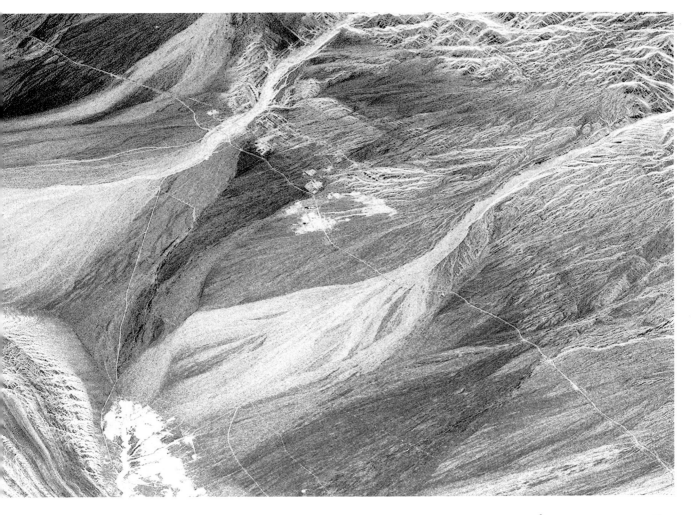

0   10km

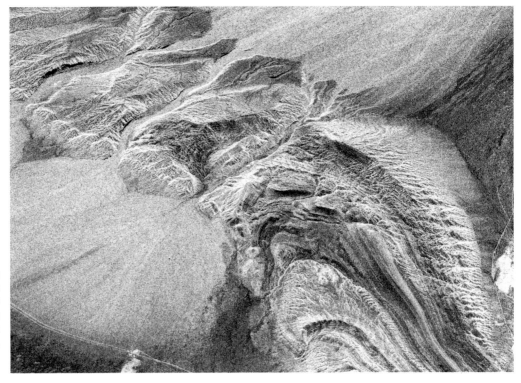

Portion of the false color SIR-C image showing different color strips indicating different rock types (R: L-HH, G: L-HV, B: C-HV).

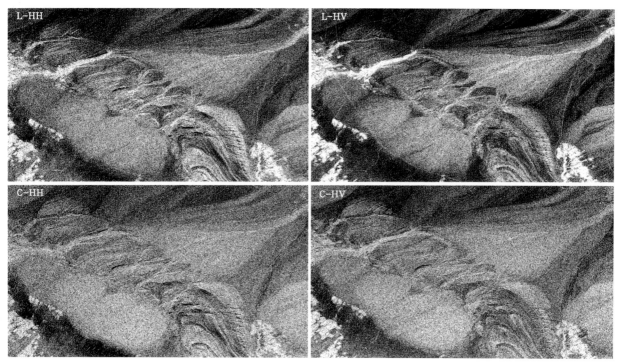

Comparison of four different black and white bands and polarization combinations of the SIR-C image for Huoyanshan Anticline.

Fold in the Western Piedmont of Taiwan

Most of the Tertiary strata in the piedmont of western Taiwan are folded into anticlines and synclines. The strikes of the folds are oriented NEE to NNE, and most of folds are asymmetric due to the forces coming from the east or southeast towards the northwest. The northwest flanks of anticlines are steeply dipped; even the strata are reversed whereas the strata of southeast flanks are slightly dipped. On the contrary, most strata of the southeast flanks are very steep or reversed. There

are many low angle thrusts in the folds, forming an imbricated faulting system.

The false color SIR-C image shows an anticline called the Chuhuangkeng Anticline, the core of which is composed of mid-Miocene sandstones and shales, and the flanks of which are sandstones and shales of late-Miocene to Pliocene.

SIR-C false color composite image showing Chuhuangkeng Anticline in Taiwan.

0   10km

Sichuan Detachment Fold

This portion of the SIR-C false color composite image shows an example of detachment folds spectacularly exposed in Sichuan Province. To the west of the image, broad open synclines form basins. The synclines are mainly composed of Jurassic, Triassic, and Permian rocks. Their stratigraphic occurrences are very gentle. An example of this structure is Changba syncline, with axis oriented towards NE (from E1 to L3) and layered structure of lithological units on the wings. From SAR image can trace to the pitching ends (such as J2, L3). The Jurassic, Triassic, and Permian rocks form scarps in some areas (e.g., J13, C6). To the east of the image, a number of triangular facets along the linear structures are very conspicuous in a densely vegetated area. The faulting system in the area indicates regional compression towards the west and northwest.

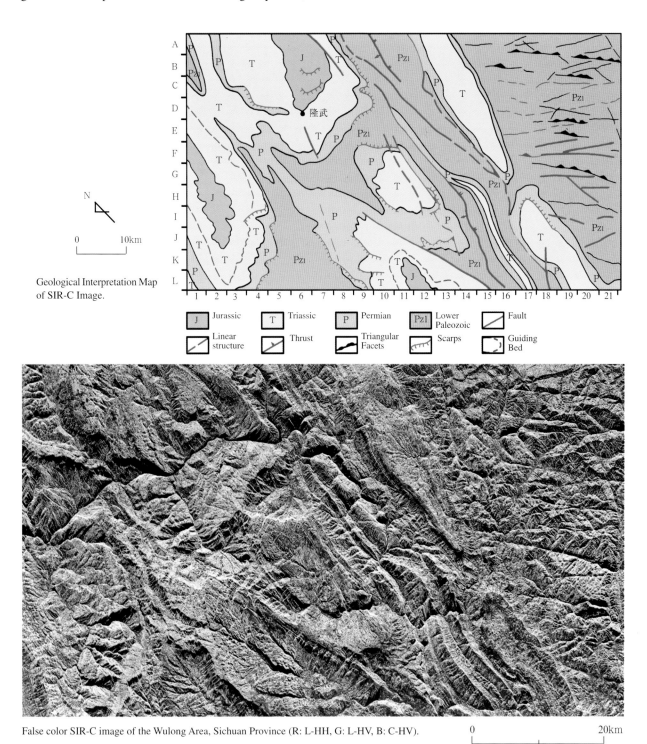

Geological Interpretation Map of SIR-C Image.

	Jurassic		Triassic		Permian		Lower Paleozoic		Fault
	Linear structure		Thrust		Triangular Facets		Scarps		Guiding Bed

False color SIR-C image of the Wulong Area, Sichuan Province (R: L-HH, G: L-HV, B: C-HV).

0   20km

131

Kuche Fold

To the north of Kuche County of Xinjiang Uygur Autonomous Region, mudstone, conglomerate and sandstone of the Kuche Formation (N_{2k}), Kangcun Formation (N_{1-2k}) and Jidike Formation (N_{1j}) dominate the outcropped rocks. There are five nearly parallel major anticlines from south to north. On the SIR-C and TM images, the above-mentioned structures are clearly seen, while the SIR-C image better shows the lithological and subtle structural information. For example, there is a layer of alluvial sediments covering Kuche Anticline and Xikuche Anticline. The structural information revealed by the SIR-C image clearly shows the bedding characteristics and anticline pitching at the east and west ends. In order to fully use TM multi-spectral information and imaging SAR information, TM bands 7, 4, 1 were first transformed from RGB to HIS; then the X-band VV polarization radar image replaced I (intensity); finally IHS to RGB were added to transform and produce an integrated image with SAR and TM information. The integrated image maintains the SAR structural information while adding TM spectral information; therefore, it is useful for geological interpretation.

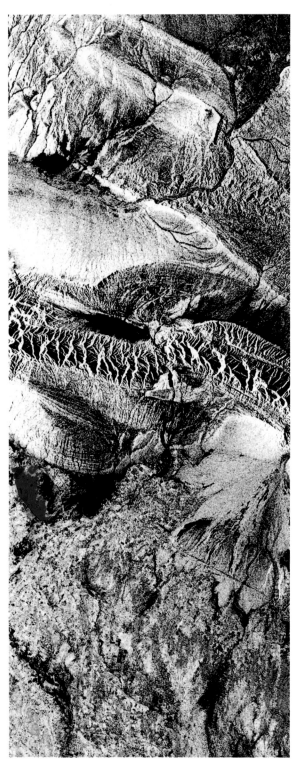

Integrated image of X-SAR and TM showing characteristics of the Kuche Fold.

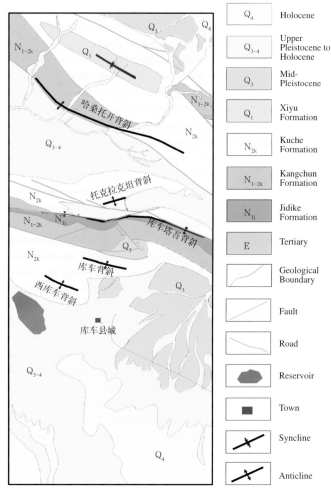

Q_4	Holocene
Q_{3-4}	Upper Pleistocene to Holocene
Q_3	Mid-Pleistocene
Q_1	Xiyu Formation
N_{2k}	Kuche Formation
N_{1-2k}	Kangchun Formation
N_{1j}	Jidike Formation
E	Tertiary
	Geological Boundary
	Fault
	Road
	Reservoir
	Town
	Syncline
	Anticline

Fold and fault structural interpretation based on the integrated image.

0   10km

False color SIR-C image showing folds in the Kuche region (R: L-HH, G: L-HV, B: C-HV).

Tiefengshan Anticline in Wanxian County, Sichuan

The Tiefengshan Anticline of Wanxian county, Sichuan Province, is a reversed anticline composed of Triassic and Jurassic rocks and has an axis oriented towards the NE. The strata sequence from core to both flanks of the anticline is the following: Badong Formation (T_{2b}) - Xujiahe Formation (T_{3xj}) - Zhenzhuchong Formation to Ziliujing Formation (J_{1-2})- Suining Formation (J_{3s}).

On the Chinese airborne L-band SAR image, the core rocks of the anticline have become relatively low hills due to their poor resistance to weathering. The lithological bedding and triangular facets are very conspicuous. The strata occurrence of the anticline changes rapidly from nearly upright at core to gently dipping along both flanks. Of course, the reversed strata bedding information can not be obtained from a remote sensing image. We can only say that they are dipped towards the NW (the occurrence of the strata in normal anticline here is dipping towards the SE).

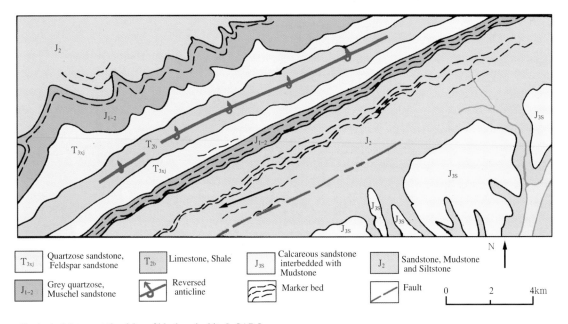

Geological Interpretation Map of black and white L-SAR Image.

Black and white L-SAR image showing the Tiefengshan Anticline of Wanxian County, Sichuang Province.

0   4km

Synclinal Structure of Binxian, Hunan

A synclinal structure in Hunan Province was interpreted according to an image taken by the second generation airborne SAR developed by the Chinese Academy of Sciences in 1983. This syncline is characterized by a bedded strata and relatively bright radar return. From the axis to both wings, the strata rocks change from hard to soft, and then to hard, exhibiting a symmetric nature. Comparative analysis of the strata indicates this syncline is pitched towards the south.

Arcuate Structures

Arcuate Structures in Northeast Zhaoqing

A portion of the Wushan-Sihui deep fault runs into this SIR-C image scene, which is situated in the Zhaoqing region of Guangdong Province. This NE oriented fault, formed during Caledonian movement, has differential expressions on different parts of the SIR-C image. In the middle section of the SIR-C image, the arcuate fault is clearly seen. To the west of this structure, the largest gold deposit in Guangdong Province occurs (outside of the image). At the conjunction point of faults in the arcuate structure, gold mineralization occurs. Local people have named this area the "gold pit." To the middle east of the SIR-C image, an oval shaped area light red in color is an area of exposed biotite granite resulting from the Yenshan movement (J_2-K_2).

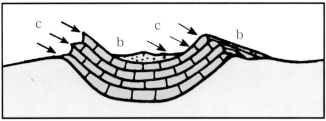

Sketch map showing radar wave interacted with syncline.

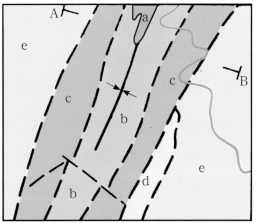

Geological Interpretation Map of SAR image.

Profile of the syncline.

Airborne SAR image of Dutou area of Binxian county, Hunan Province, acquired in 1983 by CAS/SAR.

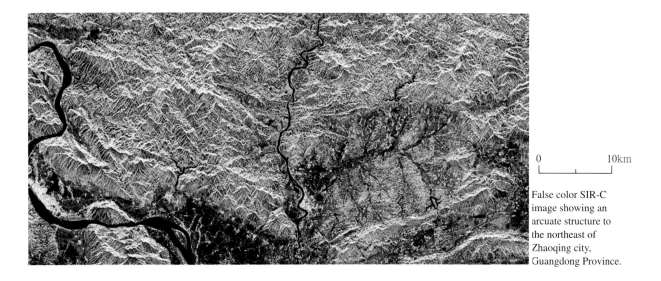

False color SIR-C image showing an arcuate structure to the northeast of Zhaoqing city, Guangdong Province.

Yeder Arcuate Structure of Altay

This SIR-C image shows an arcuate structure, which is situated, in Fuyun County of the Altai district, Xinjiang Autonomous Region. The terrain varies from gobi, rolling hills, to high mountains from south to north. The famous Erqisi River runs through the middle of the image. A gold deposit in the white circle on the east portion of the geological interpretation map is controlled by this arcuate structure, which is related to marine volcanic deposition.

Geological Interpretation Map of SIR-C image.

γ₄	Granite
γ₀₄	Plagiogranite
γδ₄	Biotite Diorite
γβ₄	Biotite Granite
δ₄	Diorite
δβ₄	Migmatite
~	Biotite Diorite
x x	Gneiss
	Fault
	Circular Structure

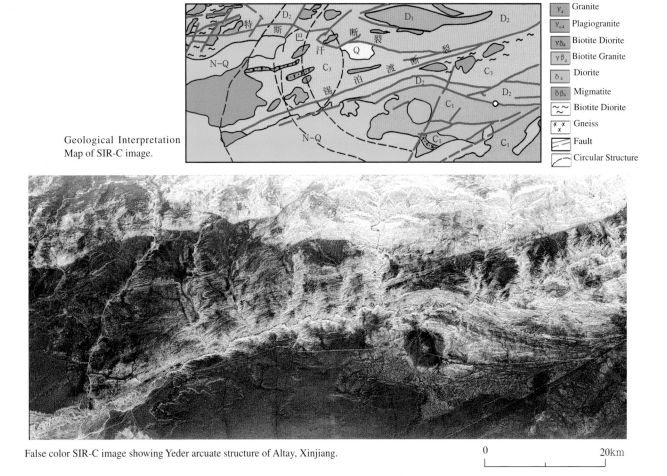

False color SIR-C image showing Yeder arcuate structure of Altay, Xinjiang.

0   20km

Luojing Arcuate Structure

The Radarsat image acquired on March 25, 1996, with center coordinates of 22.88°N latitude and 111.32°E longitude, shows an arcuate structure in the Luojing area of Guangdong Province, which reflects differential movement of the regional circular structure. North of this circular structure, circular features are not obvious on the image, and neotectonic movement is well expressed. In the middle part of the circular structure, an oval shaped depression was developed during the Yenshan movement, accompanied with a magma intrusion. A Quaternary karst basin was developed in the southern part of the circular structure, which extends along the arc towards the south. A large fault oriented NE separates the circular structure into two parts. To the north and south of the fault, different image and tectonic movement characteristics are exhibited.

There are several arcuate folds in the east and west of the circular structure, such as the Luomo Syncline, the Luojing Compound Syncline, and the Chuanbu Anticline. These folds can be identified from Radarsat image according to strata information and geometric characteristics of the SAR image.

Black and white Radarsat image showing arcuate structure in Luojing County. Guangdong Province (the bright radar feature at the lower middle of the image is the city of Luojing and at the middle right is the city of Luoding).

0   20km

LITHOLOGICAL ANALYSIS

Intrusive Rocks

Ertai Area of Altai, Xinjiang

This false color SIR-C image shows circular bodies intruded into Devonian and Carboniferous rocks. There

Meixian Area of Shanxi

On September 17, 1979, the first SAR image of China was acquired by a proto-type airborne SAR system developed by the Institute of Electronics, CAS. The SAR operated at X-band and HH polarization. From this SAR image, the geology of Meixian county was interpreted, including granite, metamorphic rock, conglomerate, and loess.

0   10km

False color SIR-C image showing the geology of Ertai area of Altai, Xinjiang (R: L-HH, G:L- HV, B: C-HV).

are two major faults in the area. To the south of the large Mayinerbo Fault is a compressive structure belt >10 km wide and oriented EW. Intrusive rocks are distributed along the fault belt. Several gold mineralization prospects have been found in this compressive structure belt. They are mainly controlled by secondary faults related to shear structure.

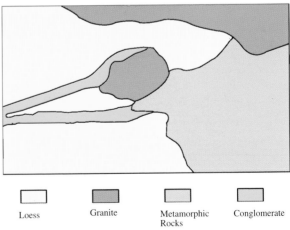

Loess   Granite   Metamorphic   Conglomerate
                          Rocks

The first SAR interpretation map of China.

China's first SAR image (taken on September 17, 1979).

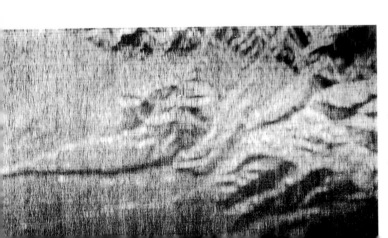

Pingshi Area of Hunan

On the SAR image covering the Pingshi to Wushui area at the boundary of Hunan and Guangdong Provinces, acquired in 1983 by the Chinese Academy of Sciences, three lithological units were interpreted. They are Upper Cretaceous (I), Upper Devonian (II), and Lower Carboniferous. This interpretation is contradicted by a unique lithological type on the geological map of the area. The tonal variation in the three rock types is also very obvious, the brightness of which is III > II > I. The possible interpretations for this are: 1) the imaging time was in rainy weather, the conglomerate (I) surface was smooth, resulting in lower radar return, or 2) there was relatively more vegetation on the surface of Carboniferous rocks, which gave rise to a rough surface and high dielectric constant, therefore producing strong radar return.

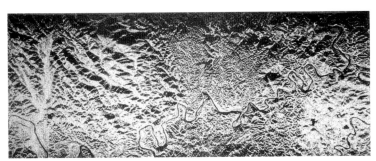

Airborne CAS/SAR image of the Pingshi area (acquired in 1983).

Geological Interpretation Map of the CAS/SAR image

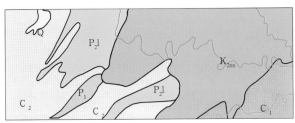

Geological map of the Pingshi area

Zhangjiakou Area

In February 1992, CAS/SAR operating in X-band HH, HV and VV polarizations and A, B modes for the first time acquired multipolarization and multiple depression angle SAR imagery. Two A mode strips of 20 km × 18 km were mosaicked and a false color image composite of three polarization channels was produced. On the false color multipolarization SAR image, Pleistocene subulous clay appears in light blue tone, which differs from other types of bedrock, e.g., intrusive rock bodies in purple. An H vector polarization gives rise to strong radar return whereas a V vector has lower radar return; therefore, these sediments appear dark on HV and VV polarizations. However, the tonal variation is similar to the tone of bedrock for which there is little tonal variation.

Multipolarization CAS/SAR false color composite image of the Zhangjiakou area, Hebei Province.

Luoding Intrusive Rocks

Luoding intrusive rock bodies of granite-porphyry are in an oval form, situated about 7 km south of Luoding city, Guangdong Province, and formed after the Cretaceous. The formation of this rock body is closely related to regional geological structures. The rock body has a small amount of dark minerals, and rock components have little variation from the margin to the center. This rock body contains more than 40% or 50% potassium feldspar. Garnet and schorlite are also present in the rock. Some gold-bearing quartz veins present in the vicinity of the rock are possibly related to the intrusion of granite-porphyry rocks.

On the GlobeSAR and Radarsat images, the Luoding granite-porphyry body appears as positive topography. Several faults and fractures are very conspicuous, indicating the intensity of recent geological activity. In comparing the GlobeSAR image to the Radrsat image, the GlobeSAR image shows much more detailed information.

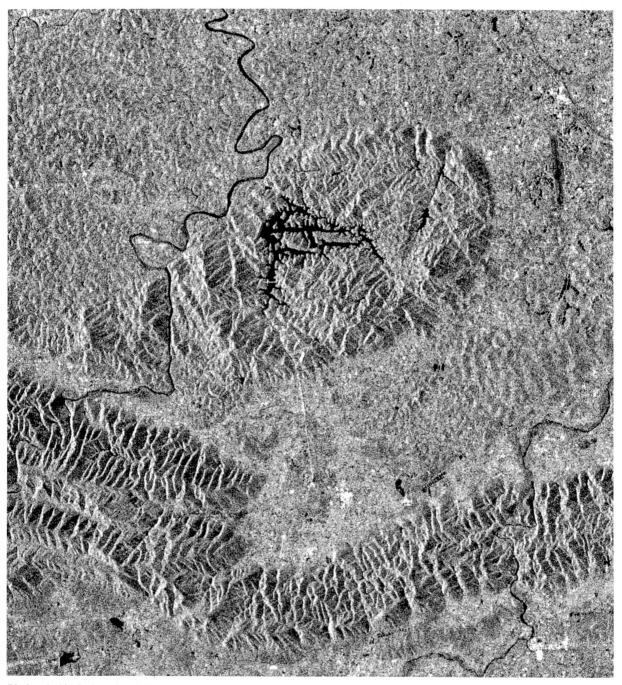

Black and white Radarsat image showing Luoding granite-porphyry body.

0   3km

3-D display of Luoding granite-porphyry body.

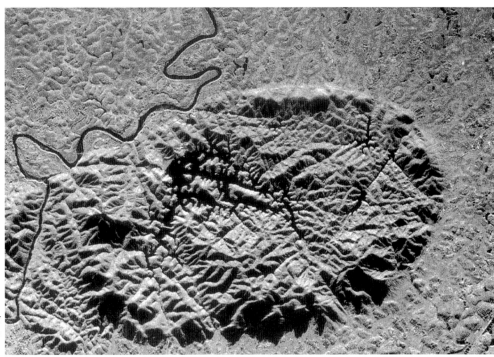

GlobeSAR image of Luoding granite-porphyry body.

Weathered Rock Basins

On the right of the false color SIR-C image of Lean County, Jiangxi Province, are two weathered rock basins of different origins. The rightmost circular basin is formed by weathering and erosion of fine to medium grain porphyritic biotite granite intruded in a late Caledonian cycle. The lageniform area to the left of the above-mentioned basin is a weathered syncline structure basin that is composed of Permian and Carboniferous rocks. On the leftmost of the false color SIR-C image, there is an arcuate syncline, the core of which is composed of Carboniferous limestone and the flanks of which are upper Devonian clastic sedimentary rocks. The syncline basin is unconformably overlain onto Cambrian regional metamorphic rocks.

Monocline Structure

The Kalpin area of Xinjiang Autonomous Region is situated in the NW margin of the Tarim Basin where the Tianshan Geosyncline joins the Tarim Platform. The Paleozoic outcrop rocks of dolomite, limestone, siltstone, sandstone, shale and other clastic sedimentary rocks constitute a series of monocline layers dipping towards the north. Rock units of Cambrian to Ordovician, Silurian, and Devonian can be identified on the SIR-A image.

The Cambrian to Ordovician rocks consist mainly of thick dolomite and limestone. There are two reasons for strong radar return from this layer: (1) SIR-A radar beam is perpendicular to the monocline structure, therefore, the corner reflector effect may be present; (2)

False color SIR-C image showing the weathered rock basins in Lean County, Jiangxin Province (R: L-HH, G: L-HV, B: C-HV).

0   20km

Black and white SIR-A image of Kalpin, Xinjiang.

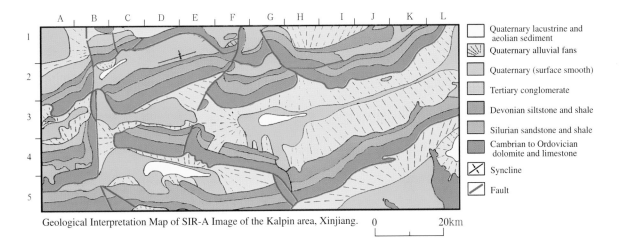

Geological Interpretation Map of SIR-A Image of the Kalpin area, Xinjiang.

0   20km

the surface is rough due to the fact that the rock is hard and maintains angular surfaces.

Silurian rocks can be easily recognized due to differential weathering of siltstone, sandstone and shale. Devonian rocks have similar rock components and similar radar return to Silurian rocks; however, this rock unit is dominated by shale. From this SIR-A image, three different types of Quaternary sediments can also be identified (see Geologic Interpretation Map).

Structural information is also very striking in the image, such as a thrust from B2 to G1/2. This area of Xinjiang is also a place where earthquakes frequently occur.

MINERALS AND PETROLEUM

Mineral Exploration

Hongguyulin Gold Deposits

An auriferous structural belt, which is about 50 km in length, has been found on the SIR-A image of Hongguyulin area of Inner Mongolia. According to gray levels, texture, and other image characters, several geological units have been mapped, which include Archaean metamorphic rock, Sinian marble, basic rock veins, Quaternary aeolian, etc. Archaean metamorphic rocks have high radar signal return due to their long evolution history of the rocks causing large lithological difference, intensive incision of the terrain, and multiple group of textures. Sinian marble mainly occurs in the north part of the study area, on the SIR-A image it appears relatively light grey. Since granitic bodies of different phases have similar physical characteristics, they are easily subjected to physical weathering, thus form relatively gentle terrain. On the SIR-A image the granitic bodies appear medium-grey tone. Some veins in the middle portion of the image (F2/F3) are very easily seen.

Although the width of individual vein is generally less than 5 m, much less than SIR-A resolution, the corner reflector effects of the veins enable them to be discerned easily.

Analysis of the SAR image suggests that the gold-bearing belt is distributed along E3 to H1. The reasons are two-fold: 1) There is an anomalous bright zone on the radar extending linearly in this area on the SIR-A image caused by the large surface roughness of a tectonically altered belt and higher dielectric constants of the altered rocks (especially mineralized rocks) than their adjacent rocks; and 2) Geological analysis of the SAR image indicates this is a gold-bearing belt since it is an important structural zone of the northern margin of the Alashan terrain which controls lithological distribution along both of its sides. Also, granitic rocks are intruded along the belt indicating that the geological environment is favorable to gold mineralization.

Analysis of the samples collected from the gold-bearing belt has confirmed that this belt is gold mineralized. Gold concentration in most altered and intensively deformed rocks is higher than 1g/t; the highest can be up to 7.94g/t, in excess of industrial required levels.

143

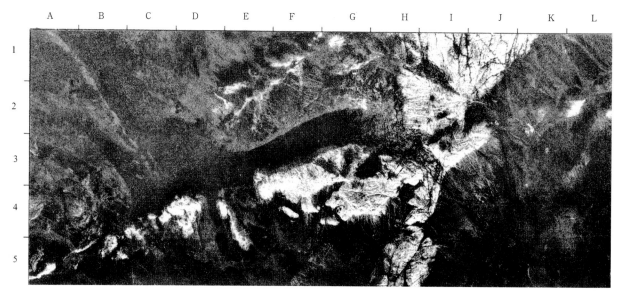

SIR-A image of the Hongguyulin area in Inner Mongolia.

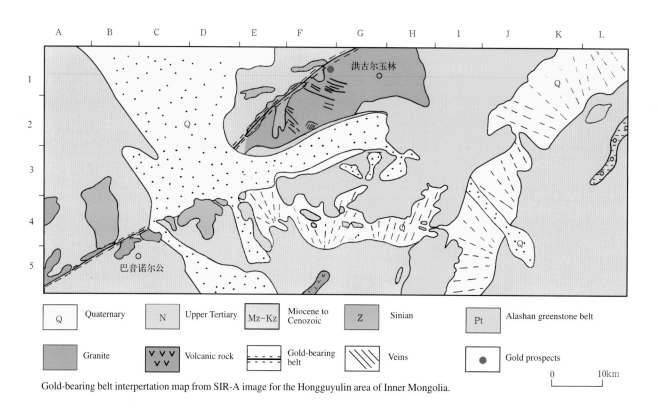

| Q | Quaternary | N | Upper Tertiary | Mz-Kz | Miocene to Cenozoic | Z | Sinian | Pt | Alashan greenstone belt |

| | Granite | ∨∨∨ ∨∨ | Volcanic rock | | Gold-bearing belt | | Veins | ● | Gold prospects |

0   10km

Gold-bearing belt interpertation map from SIR-A image for the Hongguyulin area of Inner Mongolia.

Jiaodong Gold Deposits

The Jiaodong area of Shandong Province is an important gold producing region in China. The scene of JERS-1 SAR image shows the two largest auriferous belts of Jiaodong area, Shanshandao fault zone (F1) and Jiaojia fault zone (F3). On both sides of the fault zones, tonal variations are conspicuous. The false color portion of the image is produced by integrating SAR and TM images, showing ore-bearing structures. The localities

of A and B are gold prospecting areas with gold concentrations of 1.2g/t and 1.5g/t respectively. The structure marked C is a pull-apart structure.

Mineralization in the Beishan Area, Gansu Province

The Beishan area of Gansu Province is an arid area where physical weathering is dominant. Except for the Shule and Ruoshui Rivers, which are continuously flowing streams, other rivers have become dry river

Black and white JERS-1 SAR image of the Laizhou and Zhaoyuan area of Shangdong Province. The northwest portion of the image has been merged with Landsat TM to create a false color image.

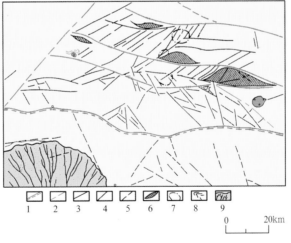

Geological Interpretation Map of the SIR-A Image for the Beishan area of Gansu Province.

1. Huge concealed fault
2. Regional concealed fault
3. Ductile shear fault zone
4. NE fault
5. Inferred fault
6. Tectosome
7. Circular structure
8. Fold
9. Yumen alluvial fan

Black and white SIR-A image of the Beishan area, Gansu Province (according to the 863-308 Report of Spaceborne SAR Applications).

145

valleys.

The SIR-A image shows Precambrian, Permian to Upper Jurassic rocks, some intrusive rocks, and unconsolidated sediments. Two shear structural belts can also be mapped from the image, the Zhongqiujing-Jinmiao ductile shear fault belt (F1) and the Xiaoxigong-Hongqiling ductile shear fault belt (F2). These shear belts are closely related to gold mineralization.

Study of the metallogenic pattern of gold in Beishan area suggests that mineralization is mainly controlled by fracture belts and ductile shear zones. Gold mineralization is associated with the secondary faults of the shear zones. In some places, host structures are in " δ " or " λ " shape fractures. The structural information extracted from the SIR-A image provides good indicators for delineating potential gold deposits (according to the 863-308 Report of Spaceborne SAR Applications).

Gold Deposit in the Habahe Area of Altai, Xinjiang

This false color SIR-C image shows the structure of gold and copper deposits in Habahe area of Altai, Xinjiang. Generally speaking, the relief in this area is high in the north and low in the south. The elevation of the mountain areas is 2500m to 3000m above the sea level. The southern area of the image is the Gobi Desert.

The area is dominated by two major faults oriented approximately NW270° to 300° and NW320° to 340°. Many intrusive rock bodies are easily identified. They mainly occur within the rolling hills and vary from basic to acidic in composition. The basic volcanic rocks intruded into the Devonian strata have been subjected to hydrothermal alteration. Within the structure zones, arcuate and secondary structures are developed. These structures are favorable places for mineralization. Most of the known gold and copper deposits in this area occur in this structure zone.

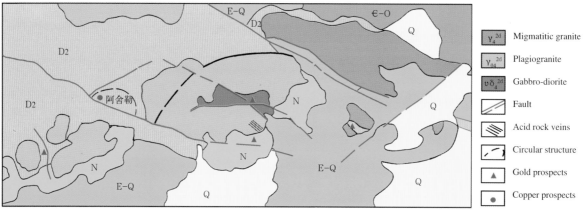

Geological interpretation map of SIR-C Image of the Habahe area, Altay, Xinjiang

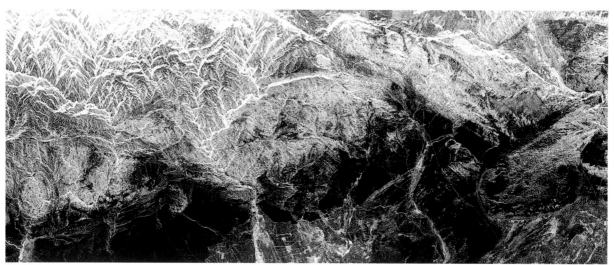

SIR-C false color composite image of Habahe area, Altay, Xinjiang (R: L-HH, G:-HV, B: C-HV).

0   20km

In late 1980s, this area was a major target for mineral exploration with emphasis on gold deposit exploration using remote sensing technology. Priori to this research work, there was no any record about the presence of a large gold deposit in this area. With the analysis to remote sensing data, a named Qiabenbulake area was considered as a gold mineralization prospect. Following the integration of remote sensing data with geophysical and geochemical data, and geological surveying and drilling, this area was proved to be a workable gold deposit. Mining activities for gold enable local people becoming rich, and improve their life.

Pegmatite Deposit in Altai, Xinjiang

The Keketuohai pegmatite deposit in the Altay area of Xinjiang is one of the largest pegmatite deposits in China. During the Caledonian and Variscan orogenies, intensive granite intrusion occurred. A large number of pegmatite veins formed in association with Variscan granite intrusions. Rare metal pegmatite veins are spatially distributed in zones around the contact zones of biotite granite and muscovite-biotite granite. There are more than 1000 pegmatite veins. Most of the veins are in lens, cupola and irregular apophyse. Structural zonation inside the pegmatite deposit is clearly seen. From the margin to the center of the cupola, the sequential pattern is as follows: Be-rich mineral belt, Be-Ni rich mineral belt, Li rich mineral belt, Ta-Hf rich mineral belt, Ta-Li-Rb-Cs-Hf rich mineral belt, and Cs rich mineral belt. These belts are distributed in concentric rings. The deposit has a very large circular open pit, which shows clearly on the Radarsat image. The bright area in the center left of the radar image is a residential and industrial area.

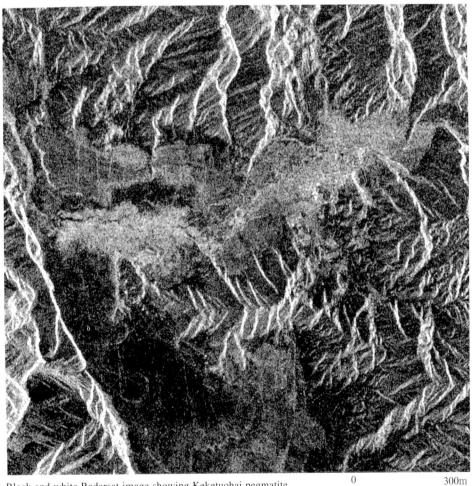

Black and white Radarsat image showing Keketuohai pegmatite deposit in the Altay area, Xinjiang.

0   300m

Petroleum Exploration

Karamay Oil Field

The Karamay oil field in Jungger Basin is one of the largest oil fields in China and has played an important role in advancing Xinjiang's economic development. The oil field has been exploited since 1950s. At present, the Karamay overthrust fault zone, which has a total area of about 5000 km², has proved to be a rare oil and gas enrichment zone. On the ERS-1 SAR image, the distribution of oil wells are clearly seen. These oil wells are point targets, resulting in strong radar-bright echoes because of metal's high dielectric constants and corner reflector effects.

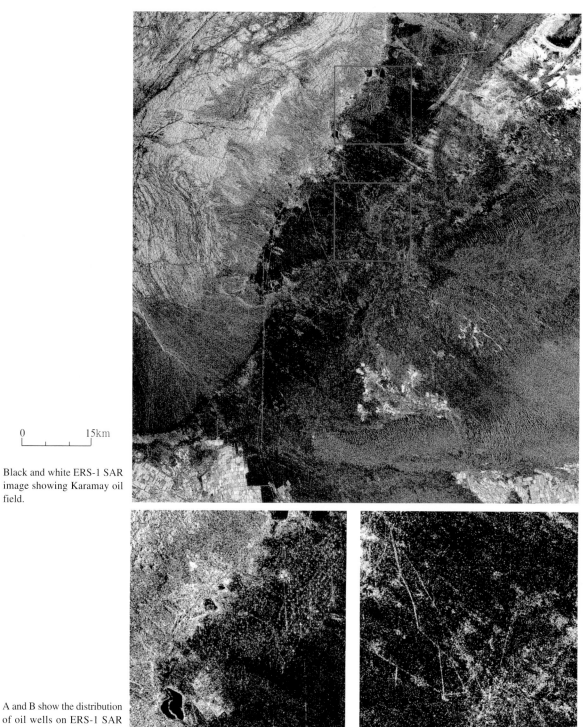

0   15km

Black and white ERS-1 SAR image showing Karamay oil field.

A and B show the distribution of oil wells on ERS-1 SAR image enlargements.

A

B

148

Shenli Oil Field

The Yellow River Delta is another oil and gas enrichment area along the coastal zone of China. The Shenli oil field, the second largest oil field in China, is situated in this area. The oil producing region is about 37,000km². They are mainly distributed in Dongying, Binzhou, Dezhou, Weifang, Zibo, Jinan, Liaocheng districts.

In Shandong Province, the favorable oil exploration areas are composed of five Paleozoic and Cenozoic subsidence and sedimentary basins. The discovered geological reserves of petroleum in the Shenli oil field is 32billion tons, and gas reserves is 257 billion cubic meters. In addition, there are about 1000 billion cubic meters of other accompanied gases. On the multi-temporal Radarsat composite image (R: July 9, 1997; G: Aug. 2, 1997; and B: Sept. 10, 1997), the distribution of oil wells and pipelines is clearly seen. (Statistical data are based on "Atlas of sustainable development of Yellow River Delta").

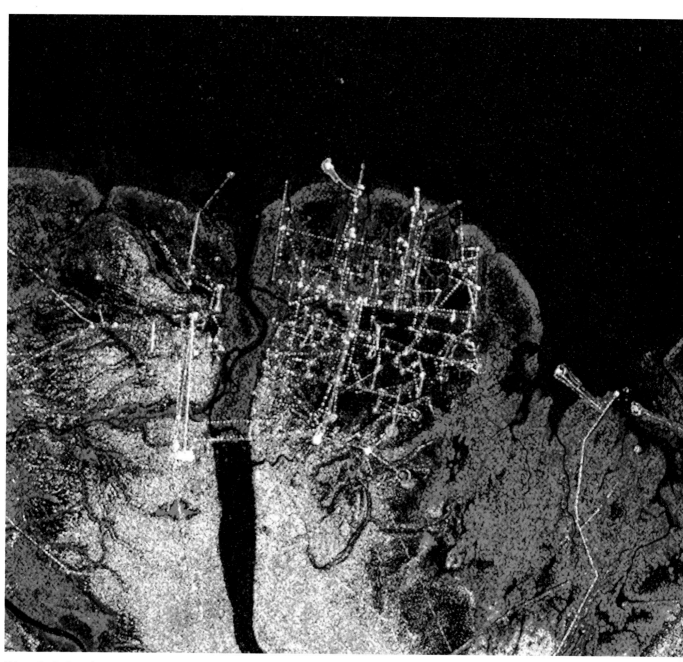

False color Radarsat image created using multi-temporal imagery covering the Shengli oil field.

0   3km

PENETRATION ANALYSIS

Penetration Analysis for Arid Area

The Alashan Plateau is one of the driest areas in China. Sand dunes are well developed and most of the rocks are buried by sand of varying thickness. Analysis of the SIR-A image has disclosed the phenomenon of radar penetration showing the detailed features of a triangular bedrock surface covered by a sand belt layer. On the SIR-A image, the portion of sand layer over the triangular rock has a bright return on radar. Several fracture structures are shown clearly. This contrasts the Landsat MSS image where the rock features underneath the sand belt are entirely invisible. This is a good example of the penetration capability of L-band SAR in an arid area. This area was chosen as a test site for SIR-C/X-SAR studies in 1994. Field investigations were conducted in this area. Digging at selected places, the field investigators found rocks underneath the sand belt at about 1m depth.

According to the geological map, the top and bottom part of the triangular rock are Precambrian metamorphic rock and Cenozoic granitoid rock; the middle part is gneiss covered by about a 2m thick sand layer. The surface temperature of the area ranges from -28° in winter to 41° in summer. The large climatic extremes cause intensive physical weathering. The windblown sands from the Badanjilin Desert smoothes the exposed rocks, but the rough surface on rocks underneath the sand layer produces a strong backscatter effect. Theoretically, to penetrate the sand layer the following conditions should be met: 1) fine grain size of surficial materials, to reduce scattering loss of radar wave energy; 2) thin surface thickness of covering sediments to enable enough radar wave energy to penetrate to bedrock; 3) dry surface and therefore low dielectric constant. All three of the radar penetration conditions at the triangular rock area have been met. The third condition is met since the annual precipitation of the area is less than 150 mm, and the amount of evaporation is 31.5 times that of precipitation.

SIR-C false color composite image (R: L-HH, G: L-HV, B: C-HV) showing the triangular bedrock feature (middle right on the image) covered by sand deposits.

Space Shuttle hand-held photography showing sand belts in Alashan desert. The red square marks the location of the triangular bedrock area shown on SIR-C/X-SAR image. The meandering river to the left of the photo is the Yellow River.

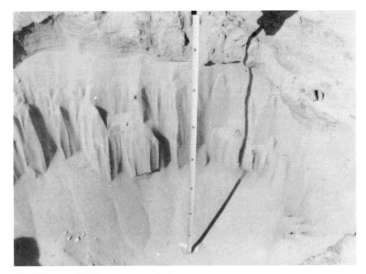

0   10km

Field Photos of SAR penetration study.

Comparison of the SIR-C/X-SAR L-HH, L-HV, C-HH, C-HV and X-VV images reveals that L-band SAR has the best penetration capability and X-band has the poorest. The L-HH image shows some rock fractures underneath the sands, but the L-HV image shows both the fractures and micro-morphology of the rock surface. The C-HH, C-HV and X-VV images show the presence of rock fractures, but they are not as clear as on the L-band SAR images.

Detection of Structural and Lithological Features Underneath Vegetation Canopy

In a subtropical region, it is very difficult to carry out geological surveying and mapping due to thick depths of soil and dense vegetation cover, poor exposure of bedrocks, and inaccessibility. Synthetic aperture radar images, especially the multi-parameter SAR, may detect geological features underneath vegetation cover. Some of the information extracted from SAR images is impossible to obtain from visible remote sensing data.

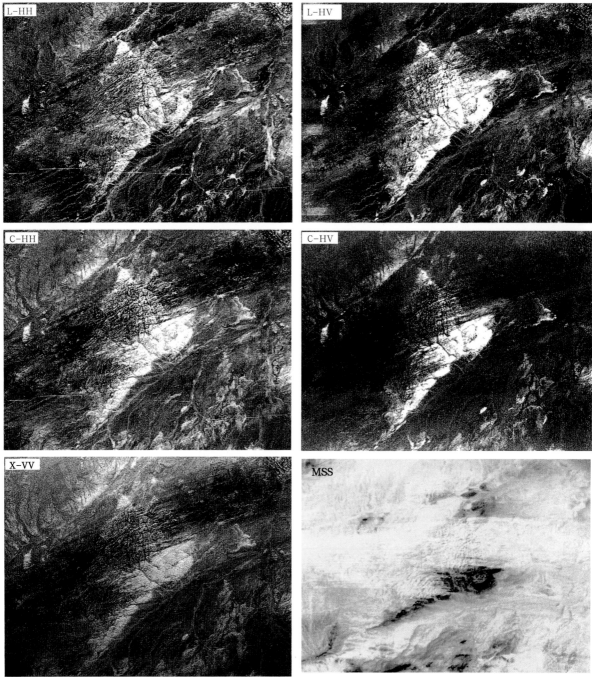

Comparison of SAR penetration ability at different bands and polarizations and their contrast to MSS image.

On April 18, 1994, the multifrequency, multipolarization shuttle imaging radar (SIR-C/X-SAR) aboard the space shuttle "Endeavor" flew over Zhao Qing region in Guangdong Province of southern China and acquired one data take at mode 11 (L-HH, L-HV, C-HH, C-HV, X-VV). The Zhao Qing test site was selected to demonstrate the advantages of multi-parameter SAR technology in detecting the geological structures and lithology underneath a vegetation canopy and to evaluate the capabilities and limitations of the SAR system for geological mapping in a subtropical region. This study was aimed at providing fundamental information for geological surveying and mapping on the scale of 1:100,000 and 1:200,000.

The test site was visited three-times for the purpose of collecting ground truth data and taking field measurements. The first trip was in November 1993 before the shuttle flight. The second trip was in April 1994 during the first mission of SIR-C/X-SAR for the purpose of collecting near real-time ground truth and relevant geological survey data and maps of the test site. The aim of the last trip was to examine, verify, and modify the interpretation results after the SIR-C/X-SAR data were available for study.

It is difficult for color infrared imagery to distinguish sedimentary rocks and determine their strike and dip orientation, which are essential elements for geological survey and mapping. The multifrequency, multipolarization color composite SAR image for Zhao Qing test site reveals the beds of sedimentary rocks underneath the dense vegetation cover. On a false color SIR-C composite image using L-HH (R), L-HV (G), C-HV (B), there are large bands of cyan, and magenta to greyish brown stripes. The stripes are parallel to each

False color composite SIR-C image showing the western part of Zhao Qing city (R: L-HH, G: C-HV, B: X-VV).

other and have clear boundaries and a sawtooth shape. These are only visible in the SAR image; no traces of the stripes are found in the Landsat TM or color infrared airphoto on a scale of 1:35,000 due to vegetation cover. The distribution pattern of these colored stripes is the same as the occurrence of sedimentary clastic rocks of the Devonian Guitou Formation (D_{2-3}) on the north slope near Zhao Qing city. These color stripes can be identified as beds of the Devonian sedimentary rocks because of the V-shaped outcrop of two adjacent formations, shown in the SAR image in a sawtooth pattern. A V-shaped outcrop of a geological boundary represents a line along which a contact between two adjacent formations or beds intersects the Earth's surface. Field work confirmed that these color stripes represent beds of sedimentary rocks of the D_{2-3} formation. There are few outcrops of these Quartzose sandstones and siltstones covered with dense vegetation growing in a thin layer of soils.

The formation of the color stripes comes mainly from the L-band image. Sun and Ranson (1995) pointed out that the backscatter received by the SAR system from a vegetated area is composed of five components. One is the trunk-ground component. L-band microwave penetrates the crown layer of the trees and reaches the ground, so there are distinct variations of intensity from the top of the slope to the bottom in the L-HH and L-HV image. It is not clear in the C-band image and there are no variations in the X-VV image. L-band SAR images show the geological features underneath a vegetation canopy better than the X-band image. The vegetation canopy is comparable to an enveloping surface over the land surface and conforming to the topography. The radar return from the X-band image comes from the top of the tree canopy whereas the longer-wavelength microwave penetrates through the vegetation layer and collects the sub-canopy information.

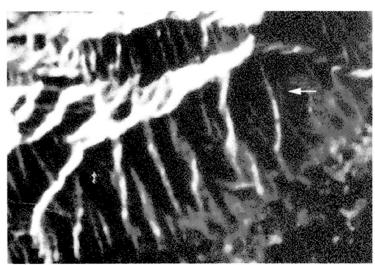

Part of the false color SIR-C image (R: L-HH, G: L-HV, B: C-HV). The arrow points to the colored stripes that reveal sedimentary beds underneath the vegetation cover.

Portion of a Landsat TM image. The arrow points to the same area as on the above SIR-C image, but there are no visible traces of lithological information underneath the vegetation.

Field photograph showing the southern slope of Zhao Qing TV tower; the arrow shows sedimentary rock beddings outcropped along a road for fire prevention.

CULTURAL FEATURES AND ARCHAEOLOGY

A city is the center of regional economy, culture and politics. Following implementation in 1980 of the Policy of Economical Innovation and Opening, the Chinese economy developed quickly. As a result, city enlargement and rural urbanization have become important issues.

Radar Remote Sensing plays an important role in city planning and development. Due to the corner reflector effect, cultural features have high backscattering capabilities when oriented parallel to the sensor, which results in a bright image tone. Therefore, it is often easy to separate cultural features such as buildings, roads, and bridges, from the surrounding environment. In macro-scale, the economical development rate and characteristics of different cities are easily delineated with spaceborne radar images according to their size and tones in the image. In addition, multi-dimensional radar images from several sensors or different system parameters (wavelength and polarization), positions, or time can be used to extract theme information for a specified city.

Radar Archaeology is a new research field in archaeology. As a result of climate change and environmental change, some ancient buildings are buried under ground. Radar's penetration ability provides a new tool for this kind of exploration.

URBANIZATION

N

0   2 km

The SIR-C image
composed by C-band,
HH and HV
polarizations in the
western Shenzhen

Shenzhen in Guangdong Province is a typical example of urbanization. The city is composed of Bao'an County and Shenzhen Economical Special Region, and the latter is divided into five districts: Nantou, Shangbu, Shekou, Shatoujiao, and Luohu. The C-band SIR-C (HH+HV) image acquired in 1994 covers the area of Bao'an County and the Districts of Shekou, Nantou, and west of Shangbu.

Xixiang is a city located in the southwest of Bao'an County, with Guang-Shen Highway crossing through. In the colorized far-infrared thermal image acquired in 1982, Xixiang is 3-kilometers away from County Bao'an and the Xixiang River crosses from north to south. At that time, buildings were limited in the south, and

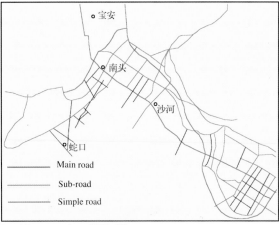

Road Interpretation of SIR-C image

0   1km

Urbanization of the town of Xixiang. The bottom left image is a SIR-C false color image acquired in 1994; the top right image is far-infrared (thermal) image acquired in 1985. The area in the rectangle in the lower left SIR-C image corresponds to the area in the bottom right thermal image.

farmland and ponds dominated the area. From the SIR-C image acquired in 1994 it can be seen that, after 12 years of development, the urban area has enlarged and connected with Bao'an County. The farmland and ponds are limited to the northeast.

The transportation map of Shenzhen Economical Special Region in 1985 displays main roads, medium roads, and simple roads plus a highway under construction. In the SIR-C image, all these levels of roads are clearly revealed. The highway has been completed, and the number of medium roads has greatly increased, forming a transportation network where simple roads are very dense. The interpretation of this image coincides with the 1985-2000 economic planning sketch map.

In 1985, the buildings are scattered in the landcover map, mainly in the center of Bao'an County, Nantou Town, Shekou Town, and Shahe Town. In the urban interpretation landcover map based on the SIR-C image,

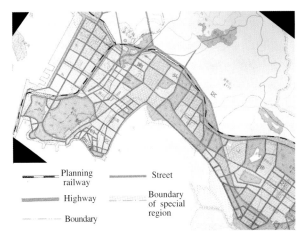

1985-2000 Shenzhen Planning Sketch Map

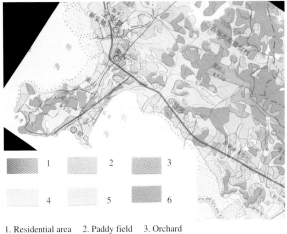

1. Residential area   2. Paddy field   3. Orchard
4. Irrigated farmland   5. Sparse forest land   6. Forest

Shenzhen Landcover Map in 1985

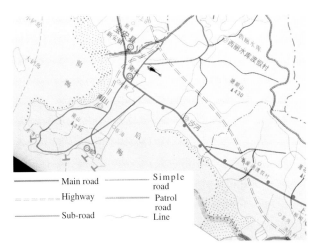

Shenzhen Transportation Map in 1985

the town centers have united, forming one large commercial/industrial area.

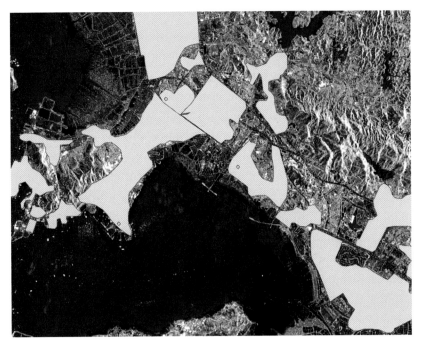

Buildings Interpretation Map of SIR-C image, where the red area represents buildings in 1994.

TRAVEL NETWORKS

City Streets

The corner reflector effect of cultural features is a unique feature of radar remote sensing. Streets and buildings or trees create dihedral corner reflectors, which produce high backscatter when their orientation is nearly parallel to the flight line. In the radar image, streets in urban areas appear as bright lines. The C-band Globe SAR image (HH+HV+VV) of Zhaoqing City was acquired on November 21, 1993. The imaging area is 21 kilometers long and 6.4 kilometers wide, and covers an urban area and part of a suburban area of Zhaoqing City. Zhaoqing is a medium size city in Guangdong Province. In the false color SIR-C image, the old blocks are obviously different from the new ones in Zhaoqing City. The old blocks are in the center of the city, characterized by narrow streets and slanted two-roofed buildings. In spite of several main streets in the old blocks, the medium streets are difficult to identify, and the simple ones are not detectable.

New blocks expand out from old ones. These new streets are long and wide, with new, high, flat-roofed buildings arranged on the sides. The main, medium and simple streets can be clearly identified in these new blocks. The main streets form the framework around the very dense network of medium and simple streets.

The information in the images from different platforms and sensors, which have different bands and spatial resolution, varies greatly. The C-band Globe SAR image acquired on November 21, 1993, has a resolution of 6 meters, so it can be used to easily classify the streets. The L- and C- band SIR-C false color image acquired in October 1993, has a resolution of 10 meters, and has less capability of being used for street interpretation. The C-band Radarsat image (H-polarization) acquired on April 1996 can be used to identify some main streets in Zhaoqing City.

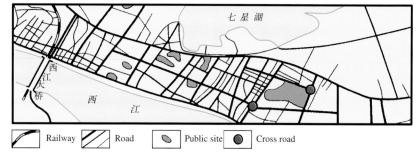

Street Interpretation Map with GlobeSAR image in Zhaoqing City.

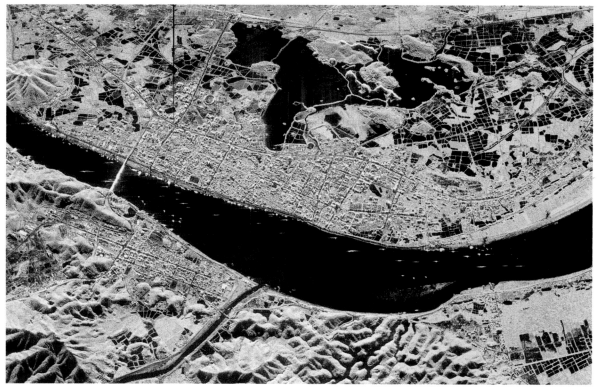

C-band GlobeSAR false color image in Zhaoqing City (R: HH, G: HV, B:VV).

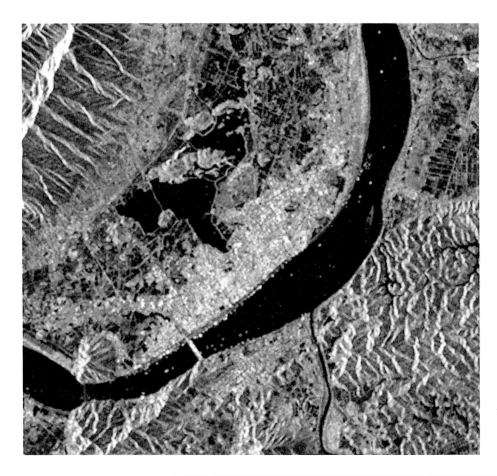

L- and C-band SIR-C false color image in Zhaoqing City (R: LHH, G: LHV, B: CHH).

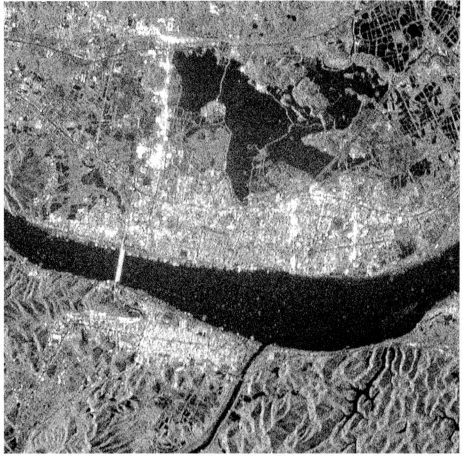

C-band multi-temporal false color Radrsat image in Zhaoqing City.

Rural Roads

The towns in South China are densely distributed, with long and wide roads and trees on both sides, so it is easy to generate corner reflection. In the L-band HH-polarization SIR-C image, Ninghe County and the large rural area around it produce high backscatter, which displays as star-studded speckles. The imaging area is 10 × 10 kilometers, with the center at Latitude 39°19'N and Longitude 117°43'E. The roads connecting counties and villages are displayed as long and bright lines. The canal is made up of concrete with trees on both sides. The water in the canal is smooth to the radar and produces very low backscatter, therefore the canal appears on the image as two parallel lines filled with a dark area. Metallic powerline towers are obvious bright spots on the image. The power lines connect the rural areas with an electric distribution center.

The roads on the image are centered at Ninghe County and radiated to the rural area. The main roads between Ninghe County and Tangshan City in the north connect the medium roads in different directions, forming a network frame. The simple roads, scattered densely in the large area of farmland, link the medium roads and form a regular cobweb-like road network.

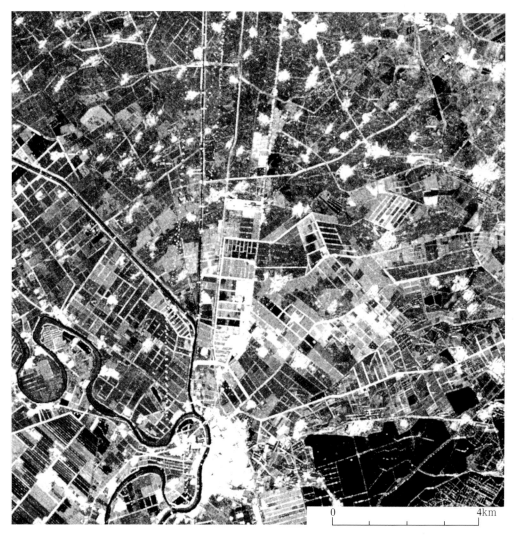

0   4km

L-band HH-polarization SIR-C image in Ninghe County.

162

IMAGING FEATURES OF CITIES

Municipal Cities

Beijing

Beijing, the capital of the People's Republic of China, is the center of national politics, economics, and culture, and the hub for national travel. Beijing

False color multi-dimensional image of Beijing City (R: JERS-1 SAR, G: TM6, B: RADARSAT).

0   10km

covers an area of 16.8 thousand square kilometers, with a population of 6,690,000 within the city limits. The major buildings of the city are centered around the Forbidden City, and expand radially. The high, flat-roofed buildings are distributed regularly, with bright tones from corner reflection helping to provide information. The Beijing Airport is located in the northeast.

The false color multi-dimensional image is made from JERS-1 acquired in 1993, TM6 in 1994, and Radarsat-1 in 1997 as R, G, and B respectively. The image center is Latitude 39°39'N, and Longitude

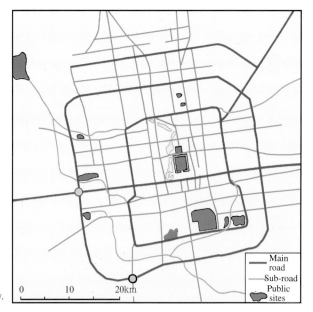

Street Interpretation Map of Beijing City.

RADARSAT image (ScanSAR mode) of Beijing City.

0   8km

116°40'E. Beijing has the typical tessellated feature of ancient Chinese capitals. The street network is organized horizontally and vertically. The vertical Central Street divides the city symmetrically into two parts. The vertical Central Street and horizontal Chang'an Street form the main cross roads at the Second, Third, and (under-construction) Fourth Ring Streets.

Tianjin

Tianjin City is located at Latitude 39°00'N and Longitude 122°30' E at the conjunction of the Hai River's five branches. Tianjin has an area of 11 thousand square kilometers and a population of 4,970,000. Most of the streets in the city are oriented perpendicular to the Hai River, with the direction of crossing at 45° with the flight line. The street information in the image is blurred.

Radarsat image (ScanSAR mode) of Tianjin City.

0   6km

Shanghai

Shanghai is located at the mouth of the Changjiang River, at approximately Latitude 31°30'N, and Longitude 121°40'E. Shanghai covers an area of 5.8 thousand kilometers, and has a population of 8,690,000 within the city limits. In the Radarsat image, the Huangpu River crosses the city from south to north, joining the Changjiang River and flowing into the East Sea. The streets in old districts are narrow and winding, so most of the buildings in the image have a low backscatter, and the street information is blurry. Airport runways can be seen on the left side of the image as long thin dark regions.

The X-band HH-polarization SLAR image was acquired on January 2, 1991, and it covers the Wusong District of Shanghai City. Buildings clearly illustrated in the image belong to Baoshan Steel Inc. At the entrance to the East Sea, boats are revealed as bright spots on the river.

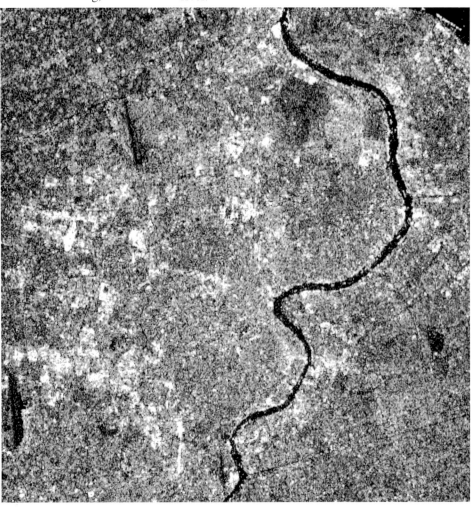

Radarsat image (ScanSAR mode) of Shanghai City.

SLAR Image of Wusong District, Shanghai City.

Hong Kong, & Taiwan's Cities

Hong Kong

The Hong Kong Special Administration Region is located at Latitude 22°32'N, Longitude 114°10'E, and is composed of Hong Kong Island, Kowloon Island, and New Territory District. Hongkong is one of the most important free-trade harbors in Asia and the Pacific Rim Area. Most of the downtown area is in the south part of Kowloon Island and the north part of Hong Kong Island.

In the false color SIR-C image (R: CHH, G: CHV, B: LHV) acquired in October 1994, the vegetated areas in the city and the surrounding mountains are displayed as gray blocks. The buildings are high and densely packed, so they exhibit high corner reflecting. Buildings oriented in different directions reveal different colors in the image. Replacing the green channel with the Landsat TM6 image in 1995 produces a better pseudo-stereoscopic effect, with the vegetated fields and mountain areas appearing as a green color. The resulting image reveals road construction on land reclaimed from the ocean. The reclaimed flat areas are displayed as a violet strip on the image. Backscatter from buildings is so bright that the street information in SIR-C image is very blurred. The three-dimensional (3-D) image of Hong Kong comes from using a second LANDSAT TM image acquired in 1994, in which the downtown area is gray.

In September 1998, the Chinese L-band Airborne SAR made an imaging flight over the Hong Kong area. In the image, we can see the New Airport and the Qing-ma Bridge. Due to the signal saturation of buildings' backscatter, the downtown area is poorly imaged. However, the image does reveal some street information.

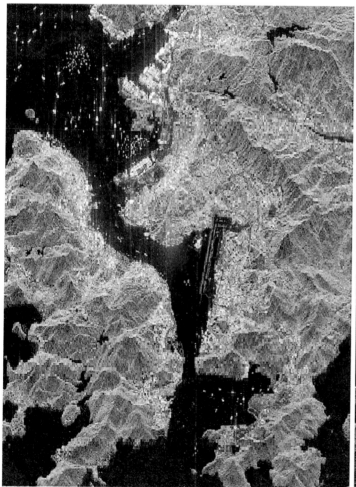 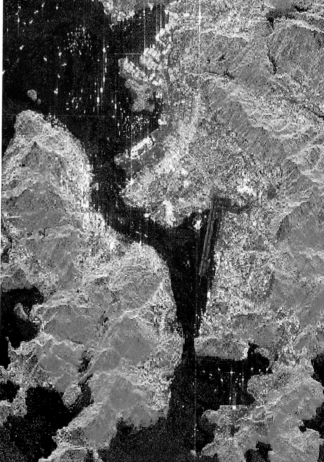

False color images composed from SIR-C and TM data of Hong Kong
(Left: R: CHH, G: CHV, B: LHV; Right: R: CHH, G: TM6, B: LHV).

0   3km

0   10km

Fine mode Radarsat
image in Hong Kong

SIR-C image of downtown area in Hongkong (left) and road distribution map in the same area (right).(Referred to "Atlas of China")

Three-dimensional (3D) image of Hong Kong.

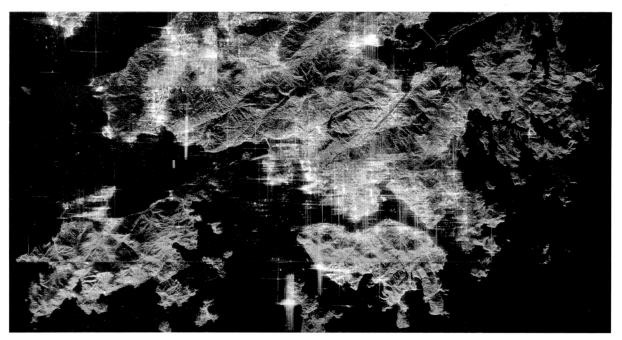

L-SAR image of Hong Kong.

Taibei

Taibei City is the capital of Taiwan Province. Taibei is located in the north of Taiwan Island at the center of the Taibei Basin and the confluence of the Danshui River and its two branches, the Jilong River and the Xindian River. The center of the SIR-C image acquired in October, 1994 is Latitude 25°02' N and Longitude 121°30' E. In this image, the gray colored river is a flood discharge canal opened in the summer season. Highways are seen as the dark lines crossing the city, the result of specular reflection. The dark linear area in the northeast of the image is the Songshan Airport. This image displays the street network of Taibei.

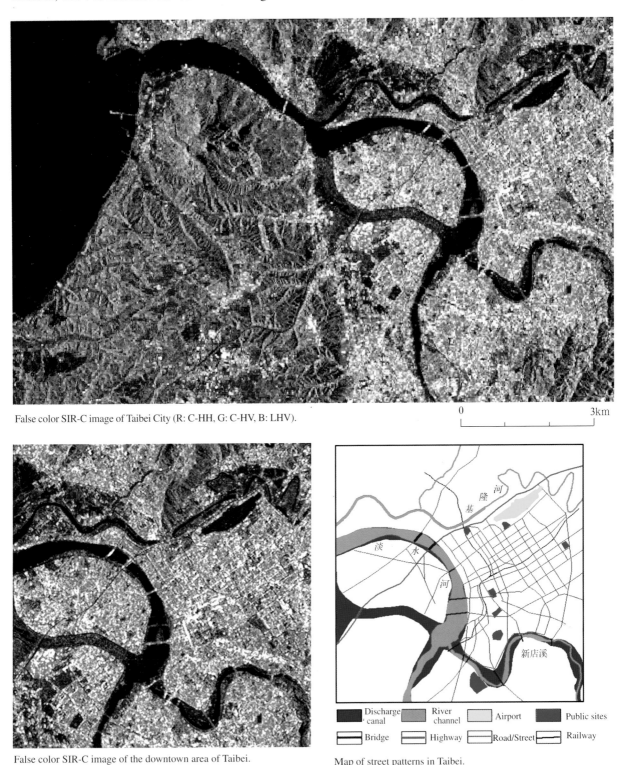

False color SIR-C image of Taibei City (R: C-HH, G: C-HV, B: LHV).

0   3km

False color SIR-C image of the downtown area of Taibei.

Map of street patterns in Taibei.

Discharge canal River channel Airport Public sites

Bridge Highway Road/Street Railway

False color SIR-C image of Taizhong area (R: L-HH, G: L-HV, B: C-HH), showing airport, downtown, and harbor areas.

0   3km

Provincial Capitals

Shijiazhuang

Shijiazhuang is the capital of Hebei Province, in the west of the Hebei Plain, at the juncture of the Jing-Guang, Shi-De, and Shi-tai railroads. With a population of 1,100,000 within the city limits, it is a typical medium size city in the north of China. The roads are long and straight, and the buildings are flat-roofed. The city has a high backscatter in the Radrsat image, with considerable street information. Villages are visible as the star-studded pattern around the city. The airport is located in the northeast of the city.

Jinan

Jinan, the capital of Shandong Province, is located on the south bank of the Yellow River, and has a population of 1,540,000 within the city limits. The street directions change in accordance with the orientation of the Yellow River, so the backscatter in the Radrsat image varies greatly, with blurred street information in the image. Daming Lake in the city center is revealed as a speckle pattern. The airport is located in the west part of the city.

Zhengzhou

Zhengzhou is the capital of Henan Province, on the south bank of the Yellow River, with a population of 1,200,000 within the city limits. Zhengzhou is located at the convergence point of the Jing-Guang and Long-Hai Railroads. In the Radarsat image, the Jing-Guang Railroad crosses the city from north to south. Land is displayed as a bright line. The airport is in the south of the city.

Hefei

Hefei, the capital of Anhui Province, has a population

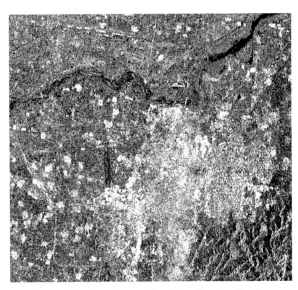

ScanSAR mode Radarsat image of Jinan City, Shandong Province.

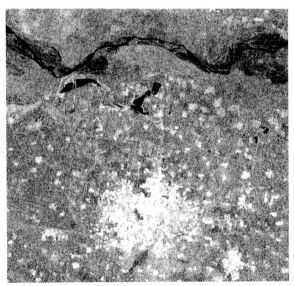

ScanSAR mode Radarsat image of Zhengzhou City, Henan Province.

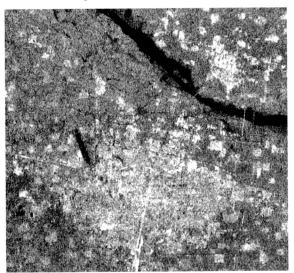

ScanSAR mode Radarsat image of Shijiazhuang City, Heibei Province.

0   10km

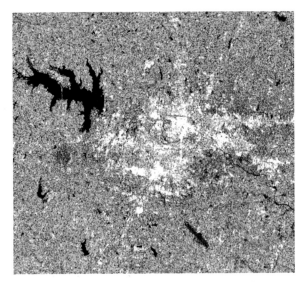

ScanSAR mode Radarsat image of Hefei City, Anhui Province.

of 790,000 within the city limits. In the Radarsat image, the Hefei River crosses the city from northwest to southeast, forming a ring-shaped moat in the city center. The road information is very blurred. The surrounding villages are sparsely scattered. The airport is located in the southeast of the city. Large reservoirs are also visible on the image.

Nanjing

Nanjing, the Capital of Jiangsu Province, is located on the east bank of the Changjiang River and has a population of 2,140,000 within the city limits. In the radarsat image, the Qinhuai River crosses the city form east to west. The dark block in the east is Xuanwu Lake. Due to changes in direction of orientation, the street information is very blurred. The villages extend along the road, forming irregular bright strips.

Wuhan

Wuhan is the capital of Hubei Province, located at the confluence of the Changjiang River and the Hanshui River. It has a population of 3,350,000. The downtown area is distributed along the banks of these two rivers. In the Radarsat image, the Wuhan Changjiang Bridge connects the north and south. The buildings have low backscatter and street information is blurred. The surrounding villages are sparsely distributed.

Guangzhou

Guangzhou is the capital of Guangdong Province, in the North of the Zhujiang Delta. In the Radarsat image, the Zhujiang River flows around the city in two paths, creating an island. The Guang-Shen Railroad crosses the north of the city and is seen as a dark line. The newly built main road is long and wide and is delineated as narrow strips of bright on both sides with dark in the middle. Due to variation in the orientation of buildings, they have a low backscatter. The towns around the city have expanded and combined with the city. The Baiyun Airport is located in the north of the city.

0   10km

ScanSAR mode Radarsat image of Nanjing City, Jiangsu Province.

ScanSAR mode Radarsat image of Wuhan City, Hubei Province.

ScanSAR mode Radarsat image of Guangzhou City, Guangdong Province.

Medium Cities and Counties

In the vast territory of China, the population density increases from west to east. The variation of geographical environment determines the difference of people's living styles, which are reflected in the Radar images.

Huairou County in Beijing

Huairou County is in the North-China Plain, at the position of Latitude 40°00'N and Longitude 116°40'E. The villages create a star-studded pattern in the SIR-C image, revealed as a large number of white regions. There are roads connecting each village. The road network converges at the downtown of Huairou County, from which it radiates in each direction.

Yuncheng City in Shanxi Province

Yuncheng City in Shanxi Province is located on the alluvial plain of the Loessal Plateau at Latitude 35°00'N and Longitude 111°00' E. The Yellow River, which meanders nearby, is the main water source for this city. The imaging time was in the low-water season, so some of the river in the SIR-C image has dried up. The population in the Loessal Plateau is much less than the in the east of China, and the roads are poorly developed. In this SIR-C image, the road information is blurred, and the villages are sparsely scattered.

Taixing County in Jiangsu Province

Taixing County in Jiangsu Province is located at Latitude 32°13'N and Longitude 120°09'E, with the Changjiang River passing through the image. The boundaries of farmlands are revealed as bright lines in

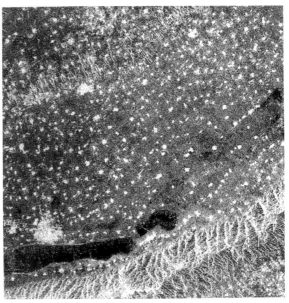

L-band HH-polarization SIR-C image of Yuncheng City in Shanxi Province.

0   3km

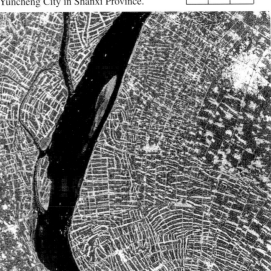

L-band HH-polarization SIR-C image of Taixing County in Jiangsu Province.

0   5km

the SIR-C image, which intercross into a dense cobweb-like network. Towns are scattered around the city. The inter-city road arteries cross visibly over the whole area, but the simple roads are blurred in the image.

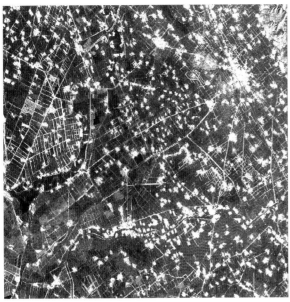

L-band HH-polarization SIR-C image of Huairou County in Beijing.

0   3km

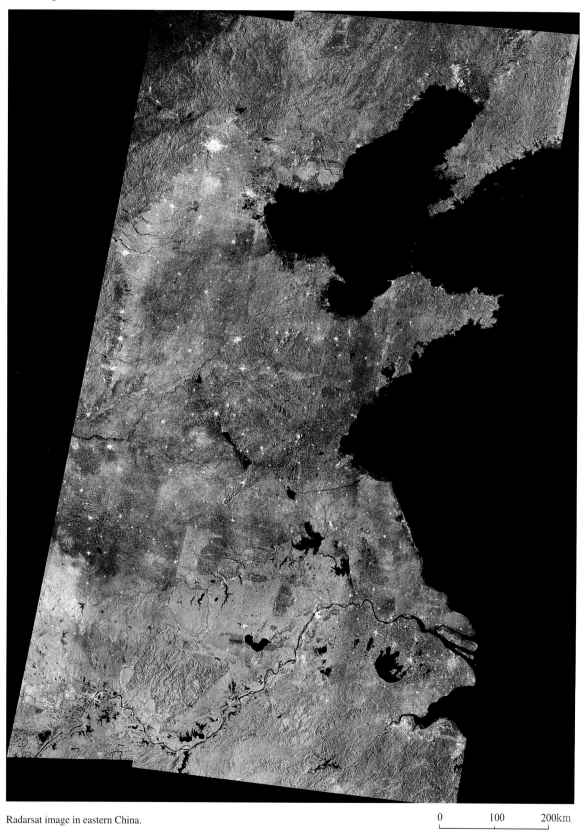

Radarsat image in eastern China.

0   100   200km

This mosaic SAR image consists of 21 scenes narrow ScanSAR mode Radarsat images acquired in May and June, 1997. Its resolution is 50m × 50m. The image shows the whole landscape from Shengyang to Jiujiang, and the areas to the east of Dabie and Taihang Mountains. The distribution of streams and rivers, and the extensive pattern of urbanization in the region are clearly exhibited in the image. This area is the most developed region with highly dense population.

DETECTION OF THE GREAT WALL

As the emblem of China, the Great Wall is one of the seven Ancient Wonders of the World. Its appearance on SIR-C image is most useful for research concerning the location and origin of the Great Wall. The images generated from different bands and different polarizations provide better information for the detection of ancient sections of the Great Wall.

Portions of the Great Wall imaged in this study are located in the northwest of China, and extend from Yanchi County in Ningxia Province to Dingbian and Anbian counties in Shanxi Province. In Yanchi County, there are three separate lines of the ancient Great Wall. The line built in the Sui Dynasty, 1400 years ago, is displayed as an orange dotted line in the image. This non-continuous dot pattern is due to the serious erosion and partial burial in the desert of this section. Two sections of the Great Wall were built in the Ming Dynasty, 500 years ago, and are displayed as two orange lines. The first wall is brighter than the second wall. The green line in the image is a forest shelterbelt along the road.

The detection of these ancient sections of the Great Wall with radar image not only allowed the Great Wall of the Ming and Sui Dynasties to be distinguished, but also allowed for differentiation of the two walls of the Ming Dynasty. In addition, radar penetrated dry sand and detected buried portions of the Great wall. The SIR-C images with different bands and polarizations have different detecting capabilities for the Great Wall. The Great Wall has a higher backscatter in HH-polarization or L-band image and a lower backscatter in C-band HV-polarization. The Great Wall of the Sui Dynasty is displayed clearly in the L-band HH image, but is almost invisible in the C-band HV polarization. The Great Wall of the Ming Dynasty is more easily detected on radar image than the Great Wall portion of the Sui Dynasty.

Field photographs of the Ancient Great Wall.

L-SAR image of the Great Wall in Yanchi County of eastern Ningxia Province.

The color SIR-C image composed by multifrequency and multipolarization data of the Great Wall from Yanchi of Ningxia to Dingbian Of Shaanxi (R: L-HH, G: L-HV, B: C-HV)

0   5km

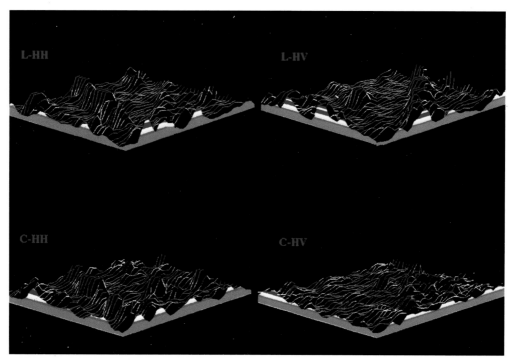

3-D view of the different bands and different polarizations on the SIR-C image.

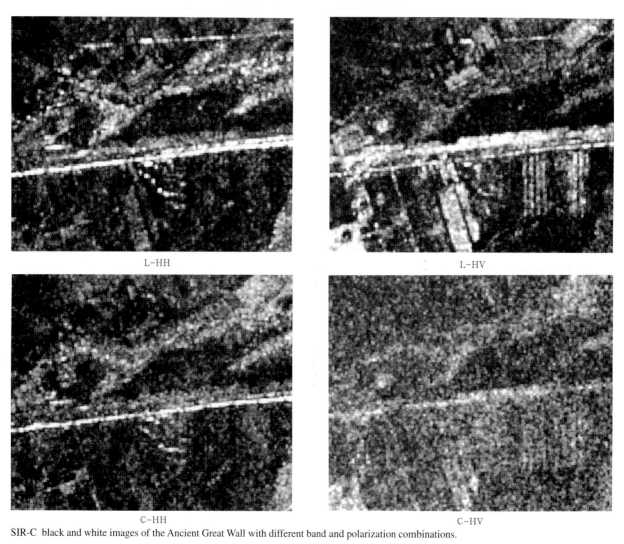

L-HH

L-HV

C-HH

C-HV

SIR-C black and white images of the Ancient Great Wall with different band and polarization combinations.

OCEANOGRAPHY

The surface area of the ocean covers over two-thirds of the Earth. The entire body of seawater is in the state of non-inertia and low-frequency movements. Scientists have assiduously researched numerous oceanic phenomena in attempts to gain more knowledge of the sea. Towards this goal, they have undertaken laborious oceanic investigations. Although the scope of this investigation is very limited considering the vastness of sea, the investigations have covered extensive time periods. It was the launch of Seasat SAR in 1978 that created a new era for oceanographers in their attempt to understand and monitor the ocean, sea ice, coastal zone processes, and oil slick movement. Spaceborne SAR can provide high resolution images of the sea; moreover, it can operate 24 hours and in nearly all-weather conditions. Following the advent of SEASAT-SAR, SIR-A, SIR-B, ERS-1/2, JERS-1, SIR-C/X-SAR and RADARSAT have been launched and operated successfully. Chinese oceanographers have utilized SAR data from these radars for oceanographic studies in China's seas. Furthermore, studies using radar remote sensing of the ocean are rapidly advancing and evolving alongside China's marine science discipline.

In the oceanography section of this atlas there are 30 spaceborne and airborne SAR images that demonstrate phenomena of ocean surface waves, internal waves, ocean and atmosphere fronts, ships and their wakes, underwater bottom topography, seasonal sea ice, oil slicks, seashore migration, and aquaculture site detection. Besides the SIR-C/X-SAR data from two flights in April and October 1994, respectively, radar data from Radarsat SAR and L-SAR have been utilized. The latter is the L-band HH airborne SAR system developed by the Institute of Electronics, Chinese Academy of Sciences. These images cover portions of the Bohai Sea, Huanghai Sea, Donghai Sea and the South China Sea as well as some coastal zones.

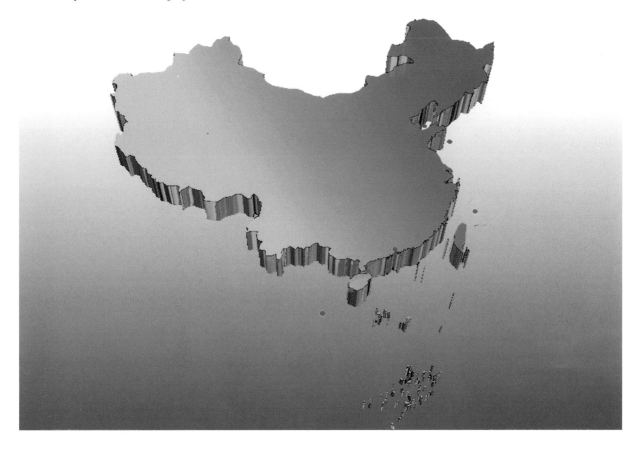

OCEAN PHENOMENA

Sea Fronts

A sea front, which is difficult to observe using normal ocean investigations, is the active water boundary between two water masses of different characteristics. Near this boundary, the gradients of current velocity, temperature, salinity and the state of the water are much larger and change suddenly forming discontinuities. These different water masses may contrast each other in their physical, chemical and even biological features. Therefore differences in back scattering exist and can often be recorded on SAR images. Similarly, for atmosphere fronts over the local sea surface, wind velocity and direction contrast sharply between two air masses of different origins and characteristics. This contrast produces a significant amount of back scattering differences and thus significant tonal contrasts on the SAR image.

The multi-dimensional false color image of Qiongzhou Strait was generated by assigning chromatic colors of red, green and blue to SIR-C L-band HV, L-band HH and C-band HH data, respectively, to data acquired in April 1994. The image reveals information on the sea front. The tidal current is flowing from top-left to bottom-right of the image. Fresh water from the rivers on both banks of the Strait is mixed with the seawater. These fronts are generated in association with the mixing processes of water masses in tidally energetic regions of Qiongzhou Strait. The image features of different colors correspond directly with the different water masses and the horizontal mixing and the vertical mixing taking place. Striations and eddies are easily visible in the right area of the Strait. Also, the sewage-discharging area of Haikou City is the tongue-like dark area adjacent to the coast and near the city.

A typical atmospheric front was imaged near Baogang, Hainan Province, in October 1994. The sea surface roughness is affected by different air masses: one cold and dry from the land and the other warm and wet from the open ocean. We can see the two fronts, indexed A and B, in the C-band image, but only front A in the L-band image. Front A is more likely an atmospheric one and B an oceanic one. The oil slick indexed C can be seen in both L and C-band images. The results of comparing L-band with C-band illustrate that C-band data is more valuable for studying atmospheric frontal patterns. Further research efforts on the mechanism of microwave imaging of fronts should be followed with simulation experiments in the laboratory and investigations in the open oceans.

It is very interesting that atmospheric fronts were identified previously in this area. They were also imaged by SIR-C/X-SAR on April 4, 1994. Again, the front E is clearer on the C-band image than on the L-band image. All these images introduced above have demonstrated important implications for detecting and understanding oceanic and atmospheric fronts.

Multi-band and multi-polarization SIR-C false-color image of Qiongzhou Strait

→ Flight   ↓ Illumination   ↗ N

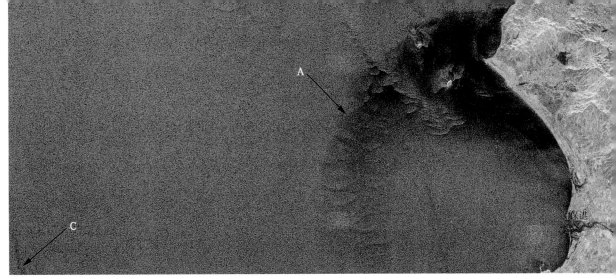

SIR-C false-color multi-dimensional image near Baogang, Hainan Province
(R: L-LL, G: L-HH, B: L-VV).

0   10km

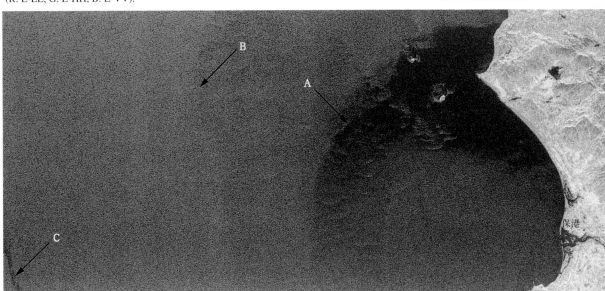

SIR-C false-color multi-dimensional image near Baogang, Hainan Province
(R: C-HH, G: Total Power, B: C-VV).

→ Flight   ↓ Illumination   ↗ N

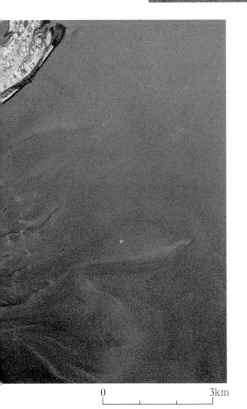

0   3km

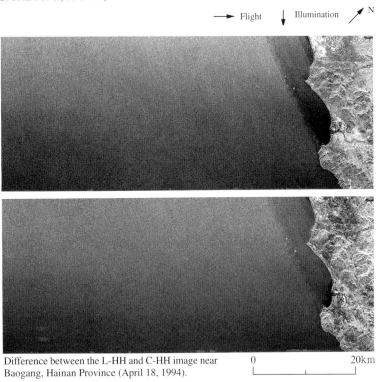

Difference between the L-HH and C-HH image near
Baogang, Hainan Province (April 18, 1994).

0   20km

183

Waves

SAR plays a crucial and significant role in studying ocean surface waves and wave climates. Some phenomena such as air-sea interaction, water masses and sea fronts, gulf streams, and cyclones are detected through the changes of wave fields. In addition, understanding and predicting wave conditions is important for harbor construction, offshore construction, operation of oil petroleum platforms, ship design, and ship routing. SAR is an important tool for obtaining wave information, day or night, from aircraft and spacecraft.

The following image of a wave field is in the South China Sea. The wind direction is southeastward, and the wave propagation direction is nearly eastward with the wavelength of about 10m.

The upper image is located near Haiyangsuo on Shandong Peninsula and was imaged on April 18, 1994 with VV polarization C-band. The image shows wave propagation and refraction. Long wavelength swells were propagating from North to South. These swells propagated onto the shadow water area and were then refracted, changing to a southeast

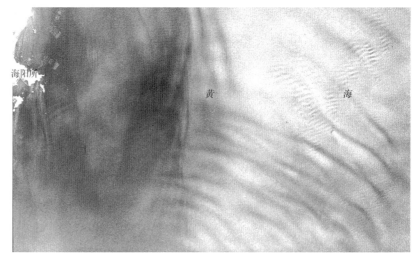

SIR-C C-VV image of waves of Huanghai Sea.

0   20km

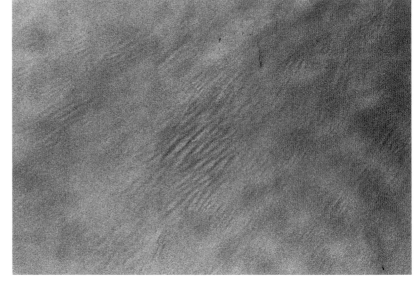

SIR-C C-VV image of swells of Huanghai Sea.

0   10km

→ Flight ↓ Illumination ↗ N

direction.

The middle image of the Huanghai Sea clearly illustrates swells propagation and the effects of local wind. The nearly westward propagating swells with a wavelength of about 1000m can be seen in most areas where the sea surface is affected by the local wind and has high surface roughness; hence, the image pattern is white-gray. But in some areas where local winds are low, there is little surface roughness; hence, the image pattern is black, and the swells are barely visible.

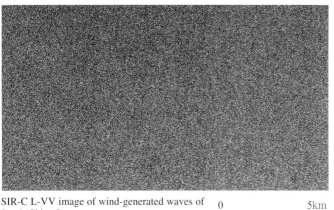

SIR-C L-VV image of wind-generated waves of South-China Sea.

0   5km

→ Flight ↓ Illumination ↗ N

Internal Waves

Internal waves occur within subsurface layers of marine waters that are stratified because of temperature and salinity variations. They are a ubiquitous and common physical phenomenon that exists not only over most continental shelf zones but also in the deep ocean. Their physical characteristics such as amplitude, propagation speed, and wavelength repeat systematically for given thermal stratification and current field conditions. Amongst the variety of internal waves, very powerful waves are created by the tide interacting with the bottom bathymetry. These waves exhibit large wavelengths and are generally called nonlinear dispersive internal waves. SAR images have being used increasingly for detecting and studying the internal waves. Internal waves are visible on radar images because they are associated with variable surface currents, which modify the surface roughness patterns via current-wave interaction. These quasi-periodic bands are commonly observed on SAR images, frequently as bright and dark bands on a gray background.

These two images of the South China Sea, which were acquired on April 18, 1994 by SIR-C, show the internal wave groupings in these areas. The main crests of group A and group B are over 80km long.

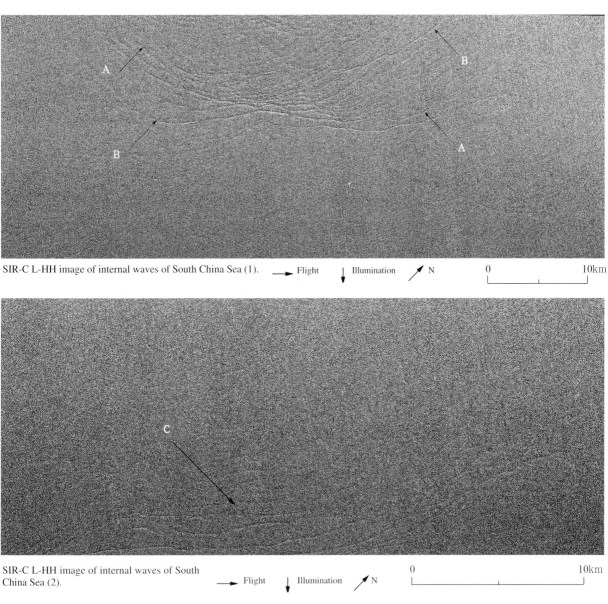

SIR-C L-HH image of internal waves of South China Sea (1). → Flight ↓ Illumination ↗ N 0 10km

SIR-C L-HH image of internal waves of South China Sea (2). → Flight ↓ Illumination ↗ N 0 10km

Underwater Bottom Topography

In the last several decades, there has been a great interest in studying the imaging mechanism of underwater bottom topography by SAR. It is now accepted that this mechanism is controlled by three processes: (1) current modulation by the bottom topography, (2) short wave spectrum modulation by the current, and (3) backscatter by the modulated short waves. These processes will make depth changes appear as contrasting variations on the image. Furthermore, the modulation of the amplitude of the sea surface waves over local underwater topography features, and then of the radar signal, is recorded on SAR imagery through the varying intensity and texture patterns. The following Radarsat image, acquired on August 2, 1997, shows underwater sandbanks in the Yellow River estuary, Shandong Province. Although underwater topography is in fact imaged, some the radar imaging mechanisms are not fully understood. The imaging conditions appear to be very crucial. For example, for the same area on the Radarsat images on July 2 and September 10, 1997, no sandbanks were imaged.

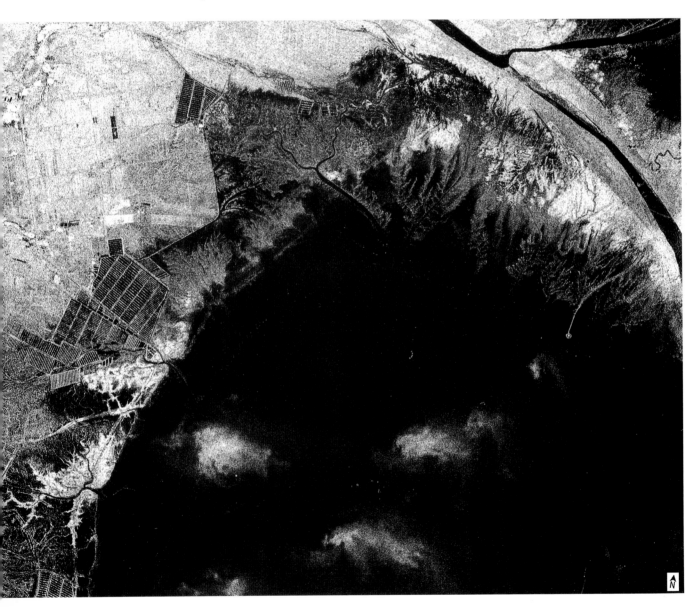

Underwater sandbanks of the Yellow River estuary.

0   10km

SHIPS AND WAKES

SAR can image both stationary and moving ships because the ships wakes have very high reflectance characteristics compared to the surrounding sea surface. Generally, the ships appear as bright point targets. A moving ship can be accompanied with a long wake. Accordingly, SAR images are used to detect and monitor ships. Studies have shown that the size, shape, type, speed, and structure of a ship can be retrieved from the SAR images in certain wave conditions.

Abundant information comes from SAR imaging of the wake. In general, the moving ship can produce both the bow wake and the stern wake and both wakes can be imaged by SAR, each having different characteristics on SAR images. The ship's stern wake appears with a relatively low radar backscatter; the increase in surface roughness at the boundary of the bow wake causes an increase in radar backscatter.

From Chinese L-SAR (L-band) image of Yantai, Shandong Province, we can see both bow and stern wakes. The bow wake appears white-gray and V-shaped, and the stern wake is a black band between both sides of the bow wake.

On the X-SAR image of the Bohai Sea, an area where there are many operating oil platforms, a large ocean petroleum transport vessel can be seen sailing through five platforms.

The difference between C-band and L-band images is obvious on both bands of SIR-C images of the Pear River estuary.

The false color multi-dimensional SIR-C image shows that the Pear River estuary, Guangdong Province, is a very busy port facility with over 200 vessels docking in-stream waiting to load or unload cargo.

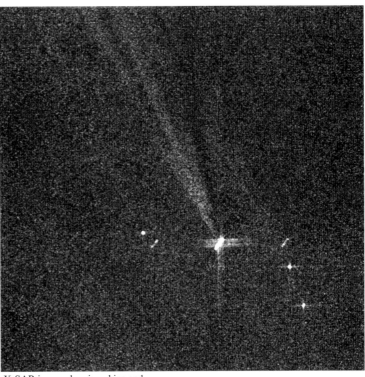

X-SAR image showing ships and their wakes in Bohai Sea surface. ↓ Illumination   0   10km

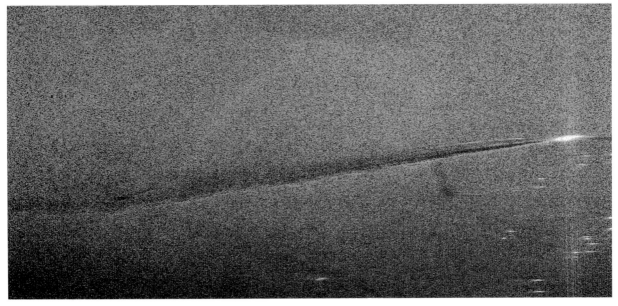

L-SAR image showing ships and their wakes near Yantai, Shandong Province.   → Flight   ↓ Illumination   ↑ N   0   1km

SIR-C false color multi-dimensional image of the Pear River estuary, Guangdong Province (R: C-HH, G: L-HH, B: C-HH).

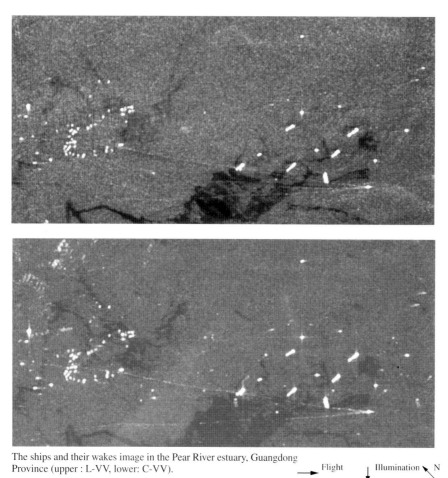

The ships and their wakes image in the Pear River estuary, Guangdong Province (upper : L-VV, lower: C-VV).

→ Flight   ↓ Illumination   ↘ N   0 _____ 2 km

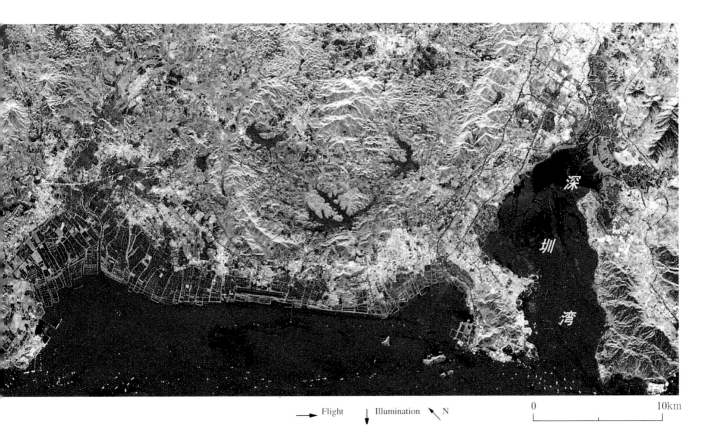

→ Flight   ↓ Illumination   ↖ N   0 _____ 10km

SEA ICE

Satellite and airborne SAR studies of sea ice started in 1978 after the launch of SEASAT-1. Since that time a series of demonstration experiments using SAR data to survey the sea ice were carried out. The main result of these demonstrations is that SAR data have proven to be a very useful source of information in operational ice monitoring. SAR data have been used to map sea ice for operational applications and climatic studies. Current and future satellite SAR systems such as ERS-2, RADARSAT, and ENVISAT operate with some tasks of identifying ice types and monitoring sea ice and iceberg movement. SAR can provide some information about ice thickness via the determination of ice type.

The Bohai Sea of China is located in the mid-latitude of the northern sphere. The seawater freezes during the long winters. Sometimes the heavy seasonal sea ices cause disastrous situations such as during the months of February and March 1969 when the sea ice caused many harbors in Bohai bay closed and the ice blocked shipping lanes. Several dozen ships were severely damaged. In order to forecast sea ice, the State Ocean Administration of China (SOA) has then directed its departments to routinely survey the whole Bohai Sea by airplane since the end of 1970s. Formerly the surveillance instrument relied on was aerial photography and infrared sensors, followed by passive microwave radiometers. Since the 1990's, SARs (airborne and spaceborne) have been incorporated into ice surveillance.

SOA has designated several ice types. These types include: (1) Fixed Ice, which is fixed to sea bottom beds in shallow water area and small islands and is produced through many freezing cycles. Fixed Ice can stretch into the open sea for a long distance. Moreover, it has strong backscattering capabilities due to accumulations of snow and materials from the land, and hence has high roughness characteristics. Fixed Ice has a definite geometric pattern that parallels the coasts in the SAR images. (2) Fresh Ice, which is produced directly through freezing seawater or snow on a sea surface with very low temperatures. Its thickness is less than 5cm. Generally, Fresh Ice forms on a very smooth sea surface. It reveals no definite geometric pattern in the SAR image. Because the surface is very smooth, it appears dark with a rough, irregular hackle boundary on SAR imagery. (3) Gray and White-Gray Ice, which are mainly produced through overlapping ice layers that were formerly frozen. The Gray Ice's thickness is greater than 10cm and the White-Gray Ice greater than 30cm. This type of ice is very destructive to ships and oil platforms. The ice layers, broken by waves and tidal currents, interweave and overlap each other forming ice ridges. Hence, the Gray and White-Gray Ice areas are very large with very high roughness characteristics so that these areas have white to gray tone on SAR images. Also, we can see clearly the leads (dark areas of open water) in these areas on SAR images. (4) Pancake Ice, which is broken ice from all types of ice. Its shape is similar to a lotus-leaf, and its size ranges from a diameter of 30cm to 300cm.

RADARSAT image of Liaodong Bay (January 16, 1997).

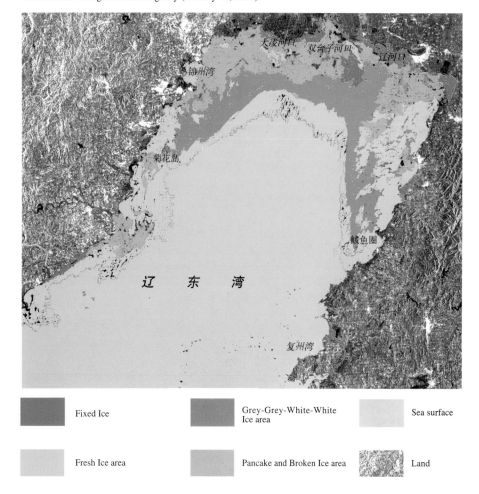

| Fixed Ice | | Grey-Grey-White-White Ice area | | Sea surface |
| Fresh Ice area | | Pancake and Broken Ice area | | Land |

Sea ice image of Liaodong Bay (January 16, 1997).

Pancake Ice can assemble over very large areas and it floats with the currents or winds.

The following Radarsat ScanSAR images and their interpretations illustrate their value for sea ice studies. These images of Liaodong Bay were acquired on January 16, January 23 and February 2, 1997, respectively. On January 16th, sea ice was located in the north area of the bay. Fixed Ice was present along the coastline and in the estuaries of the Dalinghe River, Shuangtaizihe River, Liaohe River and Fuzhouhe River. The other visible types were Pancake Ice, Gray and White-Gray Ice, and Fresh Ice. The shipping channels connecting Yingkou Harbor and Huludao Port were impeded by ice during the winter of 1997. On January 23rd, the ice edge stretched to the open sea and covered 70% of the area of the bay. Most of the harbors in this area were closed. Especially notable was that destructive Gray and White-Gray Ice was distributed over a large area. However, there was very little Fresh Ice. On February 2nd, the Gray and White-Gray Ice was broken quickly into Pancake Ice floating throughout the bay. Some Fixed Ice had begun breaking up well.

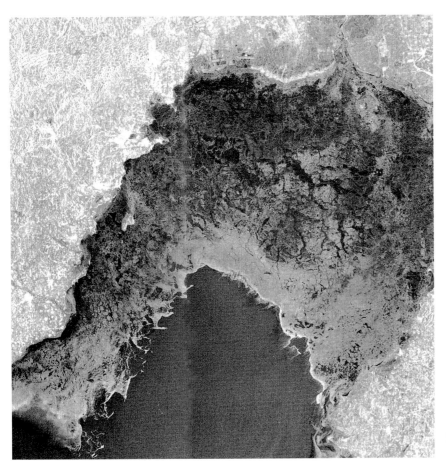

Radarsat image of Liaodong
Bay (January 23, 1997).

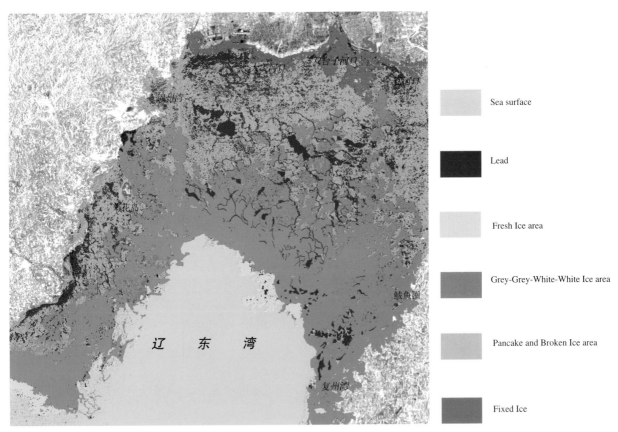

Sea surface

Lead

Fresh Ice area

Grey-Grey-White-White Ice area

Pancake and Broken Ice area

Fixed Ice

Sea ice image of Liaodong Bay (January 23, 1997).

Radarsat image of Liaodong Bay (February 2, 1997).

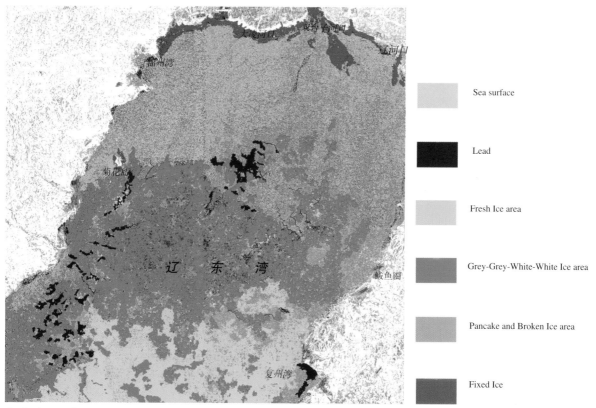

Sea ice image of Liaodong Bay (February 2, 1997).

Sea surface

Lead

Fresh Ice area

Grey-Grey-White-White Ice area

Pancake and Broken Ice area

Fixed Ice

OIL SLICKS

Qil slicks are one of source of pollution in ocean. They demanage ocean ecosystem and fishery resources. Many international pacts and laws of the country have more clausum limit severely the dimp areas. The oil slick from ships are under the surveillamce. SAR image is a useful tool for detecting illegal oil slicks. Because of their damping effects on wind-generated sea surface roughness, oil slicks are detectable on SAR image. Therefore SAR can be a tool for monitoring oil pollution of the sea by ships. Differences between a ship's oil slick and the natural oil slicks deriving from the strata of the ocean bottom can be detected by using multi-temporal SAR images. Although the presence of natural oils has been researched, we only report on ship oil slicks detected on SAR images in this section.

The following image locates in Huanghai Sea imaged by SIR-C on April 18. The center of image is 35°44′ 112″ and 122°39′ 48″. A 20km long and 100m wide oil slick strap can be detected. Also we can see the ship (B) that causes trouble was escaping southwards. Another oil slick (C) earlier the slick (A) has already diffused by the sea waves.

On the SIR-C image of Donghai Sea acquired on October 10, 1994, there are two oil slicks indicated by A and B.

There are also oil slick image patterns on the airborne L-SAR image acquired near Yantai City.

During the L-SAR experimental mission offshore Yantai City, Shansdong Province, we have the HH airborne SAR image. There is a 7km long and 30m wide oil slick strap patterns in the image. Obviously it is the ships that cause the oil slick.

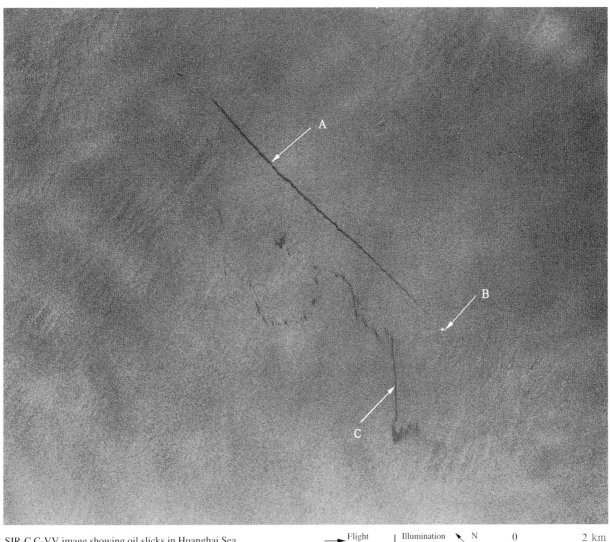

SIR-C C-VV image showing oil slicks in Huanghai Sea.

→ Flight   ↓ Illumination   ↘ N   0 ⊢—⊢ 2 km

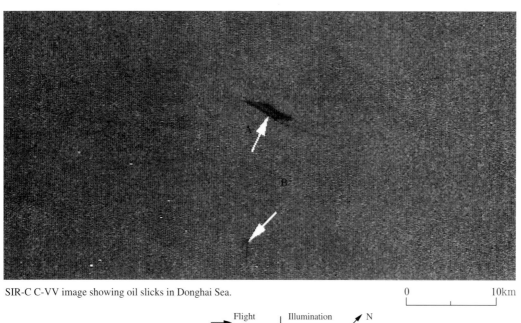

SIR-C C-VV image showing oil slicks in Donghai Sea.

0   10km

→ Flight   ↓ Illumination   ↗ N

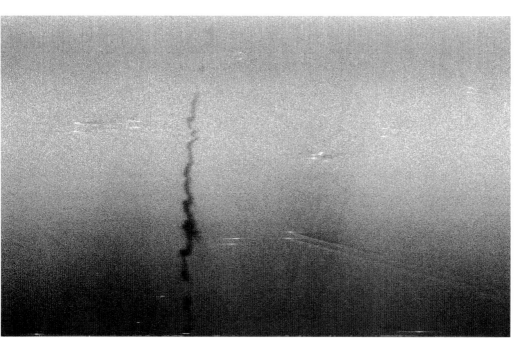

Airborne L-HH SAR image showing oil slick near Yantai, Shandong Province.

0   1km   → Flight   ↓ Illumination   ↑ N

COASTAL EVOLUTION AND AQUACULTURE SITE

Wet Land Recognition in Ninghe Area

The Ninghe area, Hebei Province is located at the center of the Bohai Depression. Migrating rivers sculptured this plain and brought a high sediment load producing a constant outward extension of the coastal area. Controlled by crust subsidence and river deposition, numerous wet lands were formed on this coastal plain. Change to wet lands over time can be an indication of coastal migration history. For human beings, wet lands are an important natural resource that needs to be utilized rationally.

Wet lands near coasts are mainly mud deposits and reed marsh, some of which have been reclaimed for fishponds or croplands. Along the coast, there are numerous salt pans and sea water fishponds. Historical documents record that a region between Ninghe and

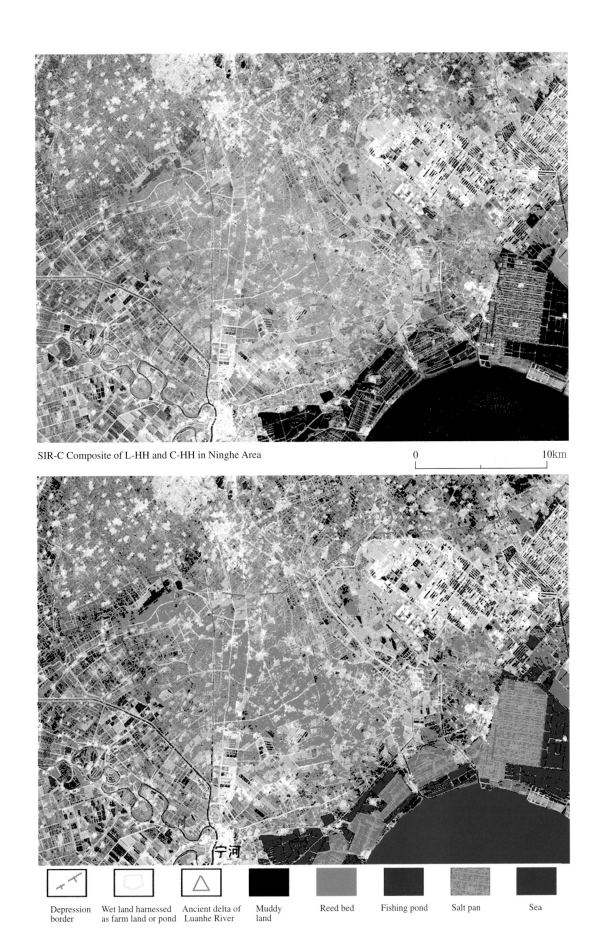

SIR-C Composite of L-HH and C-HH in Ninghe Area

0   10km

宁河

Depression border	Wet land harnessed as farm land or pond	Ancient delta of Luanhe River	Muddy land	Reed bed	Fishing pond	Salt pan	Sea

The classification map of wet lands in Ninghe Area

Tangshan was a paleo-lagoon. Before these wet lands can be transformed to croplands, the saline soil needs to go through a process of improvement. Salt-tolerant reeds are well suited for this process.

A variety of wet lands can be distinguished on the SIR-C composite image of L-HH and C-HH. Reeds yield strong responses on C-HH imagery and a weak response on the L-HH, forming an obvious contrast to other targets. Mud flats have a slightly stronger response in both channels than water bodies. Salt pans and reclaimed wet land can be recognized by their distinctive shape and location. The wet land classification map is derived from both a maximum likelihood classification and manual interpretation. The border of the Bohai Depression and an ancient delta of the Luanhe River can also be delineated on this image.

Sea Water Intrusion and River Migration in Longkou Area

Longkou County is located in the northwest of the Jiaodong Peninsula, south of which is a low hill area and north of which is an alluvial plain, the Huangxian Plain. The Huangshui River and Zhongcun River are the two main rivers in this area, both characterized by a short distance to the headwater and shallow stream depth. A great disparity in water depth and rate of discharge prevails for these two rivers between the low-water and high-water seasons.

Due to rapid economic development, the industrial and agricultural demands for water are increasing exponentially. Large amounts of ground water are drawn out each year, causing a constant reduction of ground water level and increasing severity of sea water intrusion. In the early stages, finger-like intrusions along the river channel or abandoned river channel and point-like intrusions at some industrial spots occurred. This has now expanded to a 3~5km-wide region along the coast where inland sea water has occurred to varying extents. In areas with higher ground water level, soil salinization has occurred. For areas with a ground water level of 2~4 m, salinization was caused when the salty sea water was pumped out for irrigation. False color multi-temporal JERS-1 imagery shows these human induced saline

Composite of Multi-temporal JERS-1 images, R: Jan.7 1993, G: Jul.29 1994, B: Mar. 6 1995 (red polygon indicates the waste land caused by seawater intrusion).

0   5km

Composite of SIR-C image and JERS-1 image, R: SIR-C L-VV, G: SIR-C C-VV, B: Sum of three JERS-1 images of different date.

0   5km

waste lands which exhibit a low response intensity on all dates.

From the Cenozoic Era until now, the rate of uplift of the eastern part of this area has been greater than that of the western part, forming tilting topography. This has caused the river channel to swing from east to west. Field investigation data indicates some points where abandoned channels existed. Combining these field observations with analysis of small water bodies and soil type information revealed by the JERS-1 image and SIR-C image, some abandoned channels can be delineated.

In order to control sea water intrusion, local water conservancy authority organized people to build underground dam to prevent sea water from intruding to inland and in the meantime to prevent inland fresh groundwater from flowing into the sea. Along the first gate of Huangshui River, a 6-kilometer-long underground dam has been built proved to be effective for controlling seawater intrusion.

There is a large amount of estuary delta sediment near Ganglun, where the coastline is protruding, so it can be inferred that this place that locates to the east of current Zhongcun River may have been the estuary of zhongcun River

Coastal Migration of the Pearl River Delta

The Pearl River Delta is a complex delta composed of the Xijiang River Delta, the Beijiang River Delta and the Dongjiang River Delta, formed in a paleo-bay with an area of 30,000 square kilometers. This area has been famous for its agricultural productivity for more than 3,000 years, and is referred to as "the land of fish and rice". In the past two decades, rapid industrial development has been paralleled by rapid urbanization. Consequently, the land use in this coastal environment has been highly impacted by human activities.

On the SIR-C false color composite image of L-VV and C-VV, trees and high-stem crops show up as a red color and paddy fields show up as bright cyan color due to the strong Bragg resonance response from the C-band radar wave. On wide-mode Radarsat imagery, the dense ponds in the Jiangmen and Shunde areas are extremely visible as a distinct texture. Paddy fields have a grey tone on the image. Banana fields have a white tone as do residential areas, but they can be distinguished by their different shape.

There are two main trends of land use change in this area. The area devoted to croplands is decreasing rapidly. Among agricultural land use, the area in paddy fields is decreasing, but the area in cash crop and aquatics is increasing. The main trend of coastal change is that reclaimed land is increasing. Reclaimed lands are being utilized as fishponds and not as salt pans or sugarcane fields as was the situation a decade ago. This adjustment is controlled by supply and demand of the market.

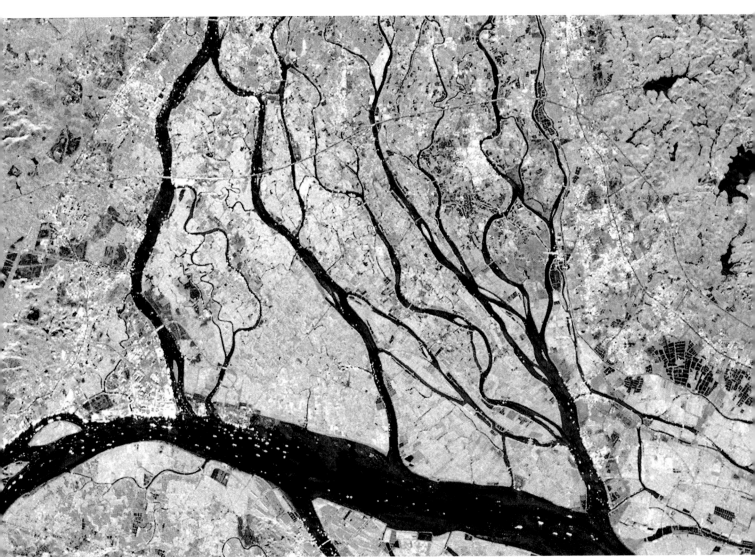

SIR-C image of Pearl river estuary

0   4km

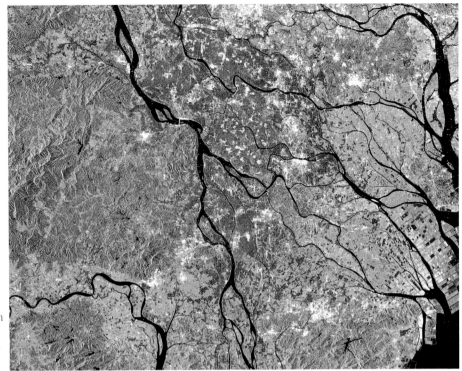

0   20km

Wide mode Radarsat image of
Pearl river estuary

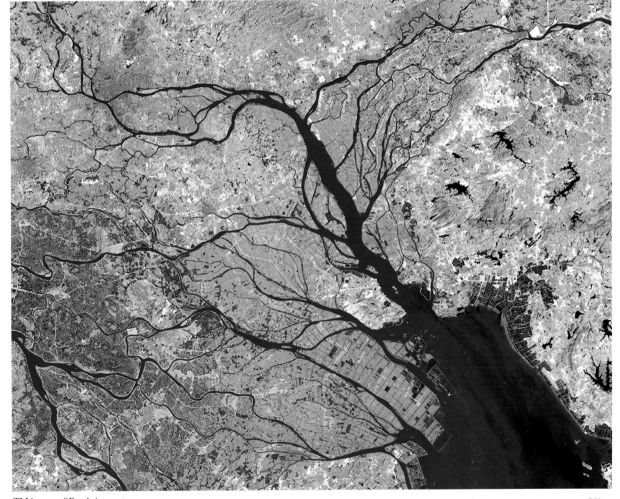

TM image of Pearl river estuary   0   20km

Aquaculture Site Monitoring in Changshan Island

Changdao Country is located near Yantai, in Shandong Province and is comprised of the Changshan archipelago. Changdao Country is a very important aquaculture base for China. The following L-SAR image covers part of the archipelago. Clearly identifiable are the distributions of aquaculture sites on the sea surface.

Also we can calculate the area and density of aquaculture site. Given the physical oceanographic parameters, we could give the manageable optimum area and density in these sites and avoid some unfit distribution.

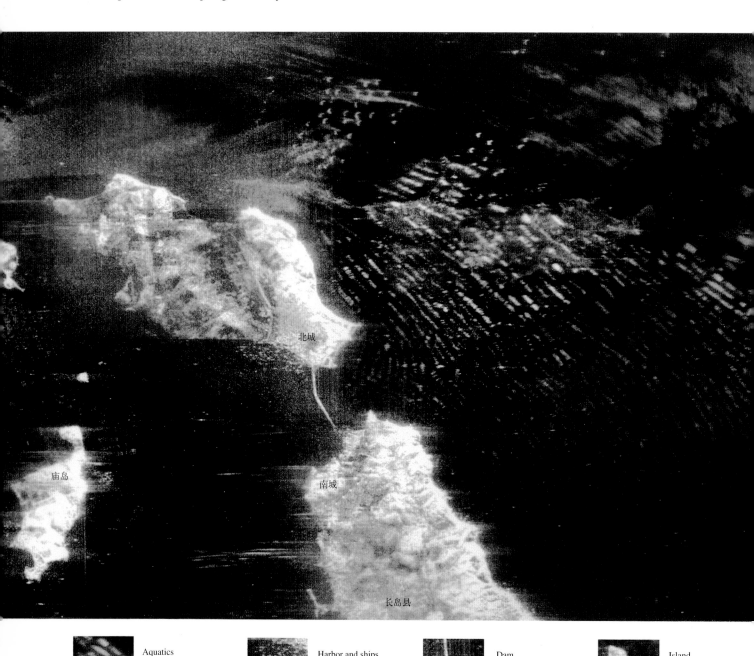

北城

庙岛   南城

长岛县

Aquatics   Harbor and ships   Dam   Island

Airborne L-SAR image of aquatics of Changdao country, Shandong Province.

0   2km

NATURAL DISASTER

Radar remote sensing technology plays a very important role in monitoring disasters, especially floods, and in assessing disaster damage. Even in some severe weather conditions, radar remote sensing technology is applicable. Radar can see through clouds and fog; delineate the spatial distribution of flooded areas; and provide information for monitoring earthquakes, landslides, mudflows, and other geological hazards.

During the summer of 1998, a catastrophic flood occurred in the midstream catchment of the Yangtze River as well as the Nen River and Songhua River basins in northeast China. Concerned authorities in China organized researchers to make full use of airborne and spaceborne SAR technology for monitoring the flood situation and estimating the extent of the disaster. Scientific data was promptly provided to the central government for fighting the flood and providing disaster relief. The Institute of Electronics, CAS, and Remote Sensing Satellite Ground Receiving Station, CAS provided the following L-SAR and Radarsat data. Flood disaster evaluations were completed by the Institute of Remote Sensing Applications (IRSA), CAS.

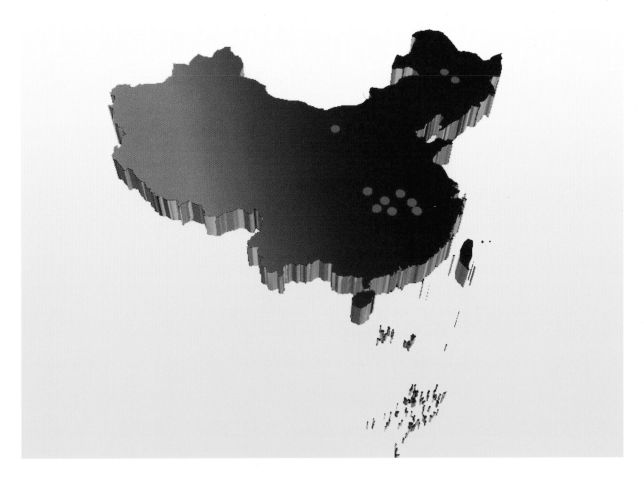

FLOOD DISASTER

Middle Reaches of the Yangtze River

The Chinese airborne L-band SAR system, developed by the Institute of Electronics, CAS, has a 3 × 3 m spatial resolution and is flown onboard a Citation S/II owned by IRSA. The system is flexible and can be used to monitor flood and other natural disasters. When the third flood peak passed through Jiu

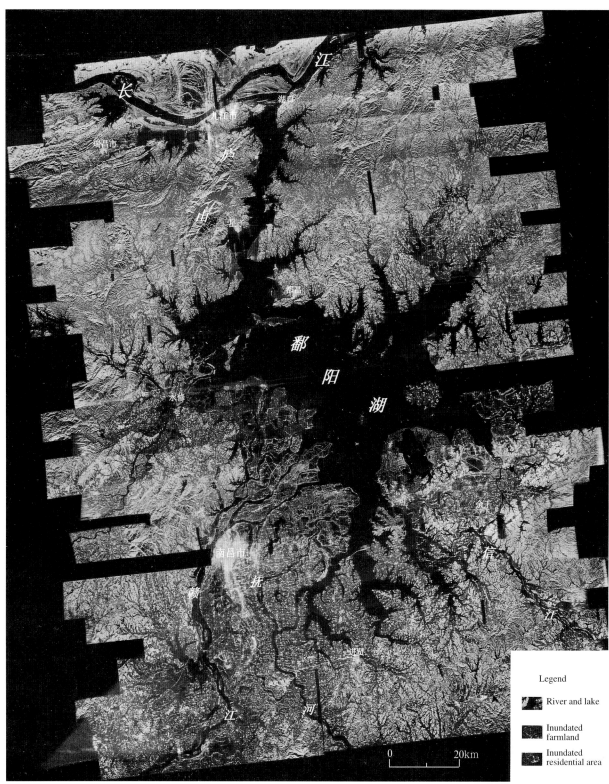

Legend

River and lake

Inundated farmland

Inundated residential area

0   20km

L-SAR mosaic image of the Poyang Lake area.

Jiang City of Jiangxi Province, the L-SAR system was flown over the area acquiring 27,700 km² real-time, high resolution data for the Poyang Lake area, which experienced the highest flood stage ever recorded. SAR images of the Dongting Lake area were also acquired during July 27 to 31, 1998. Researchers at IRSA promptly processed and moasicked the SAR images. On the basis of human-computer interactive analysis, the total flooded area and the individual flooded areas by county (city) were calculated in detail. The inundated area was further classified into farmland, town, residential area, grassland and damaged fish pond. Maps

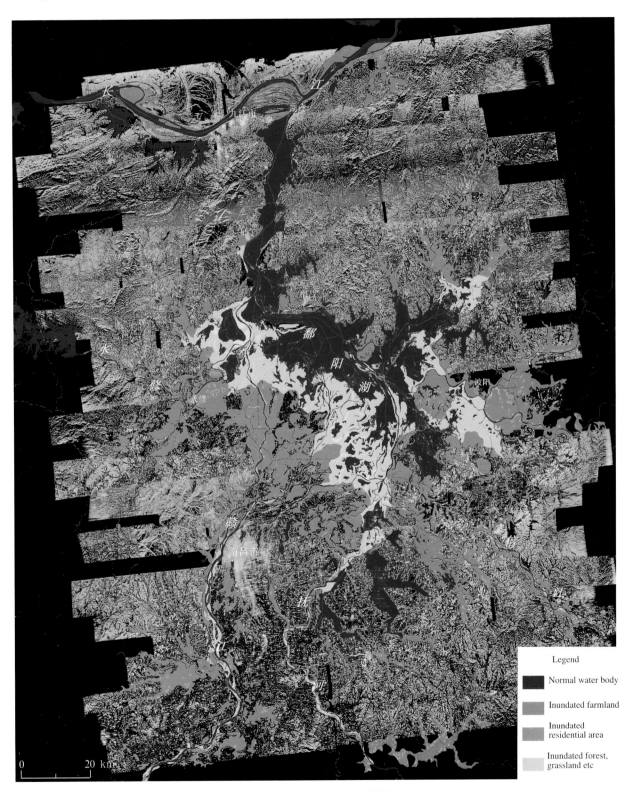

Legend

- Normal water body
- Inundated farmland
- Inundated residential area
- Inundated forest, grassland etc

Flood inundated area distribution map of the Poyang Lake region (produced with L-SAR image).

for inundated areas were produced and provided to decision-making authorities.Spaceborne SAR is superior for rapid and large-scale monitoring of flooded areas. With Radarsat data repeated every 2 or 3 days for the same area, the extent of the flood and the flood damage for a large area was obtained in a very short time.

Flood inundated area distribution map of the Poyang Lake region (produced with Radarsat image of August 25, 1998).

L-SAR mosaic image of the Dongting Lake area (data acquired during July 27 to 31, 1998; imaged area 37,000 km²).

Legend

- River and lake
- Inundated farmland
- Inundated residential area

0   20km

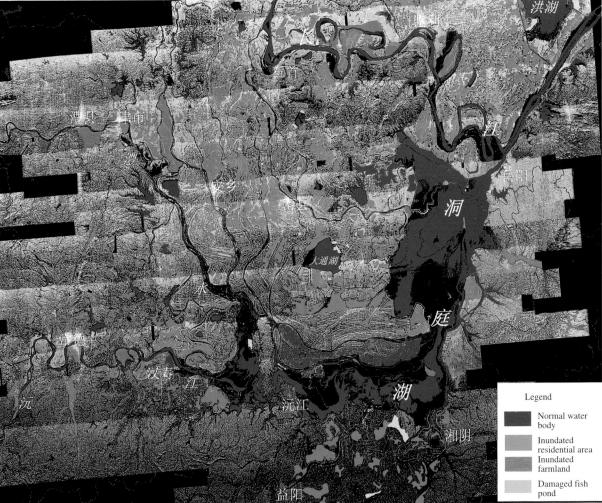

Legend

- Normal water body
- Inundated residential area
- Inundated farmland
- Damaged fish pond

Flood inundated area distribution map of the Dongting Lake area (produced from L-SAR image).

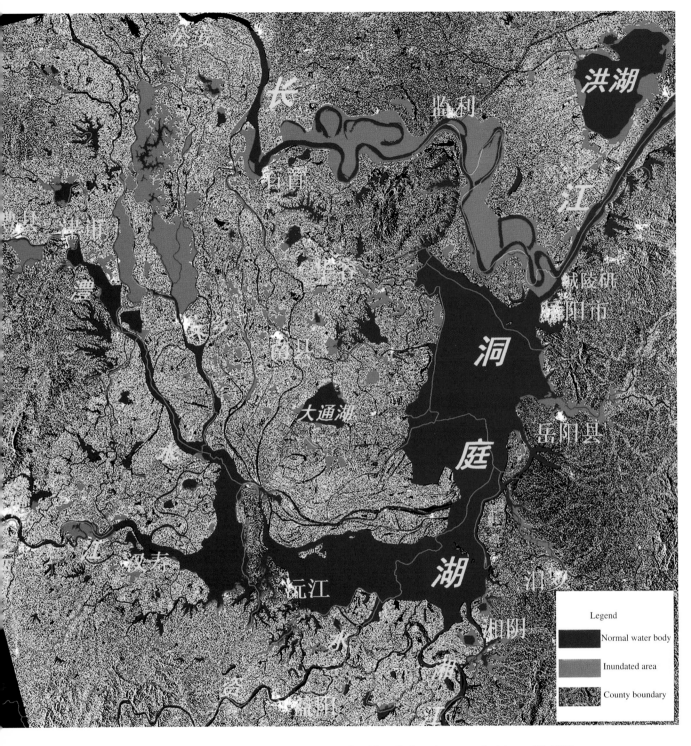

Flood inundated area distribution map of Dongting Lake area (produced from Radarsat image acquired August 27, 1998).

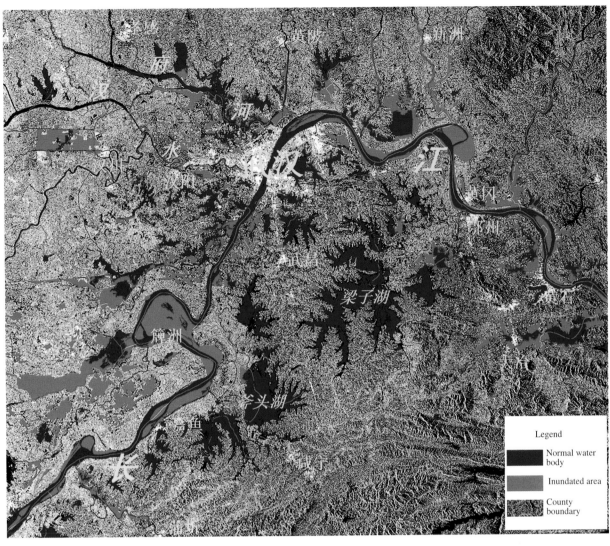

Flood inundated area distribution map of Wuhan and surrounding area (produced from Radarsat image acquired in August 25, 1998).

Damaged cofferdams of Anzhao Yuan of An'Xiang County, Hunan Province(interpreted from L-SAR image acquired in July 28, 1998).

L-SAR image acquired after dams burst in the areas of Lixian and Jinshi, Hunan Province (interpreted from L-SAR image acquired in July 28, 1998).

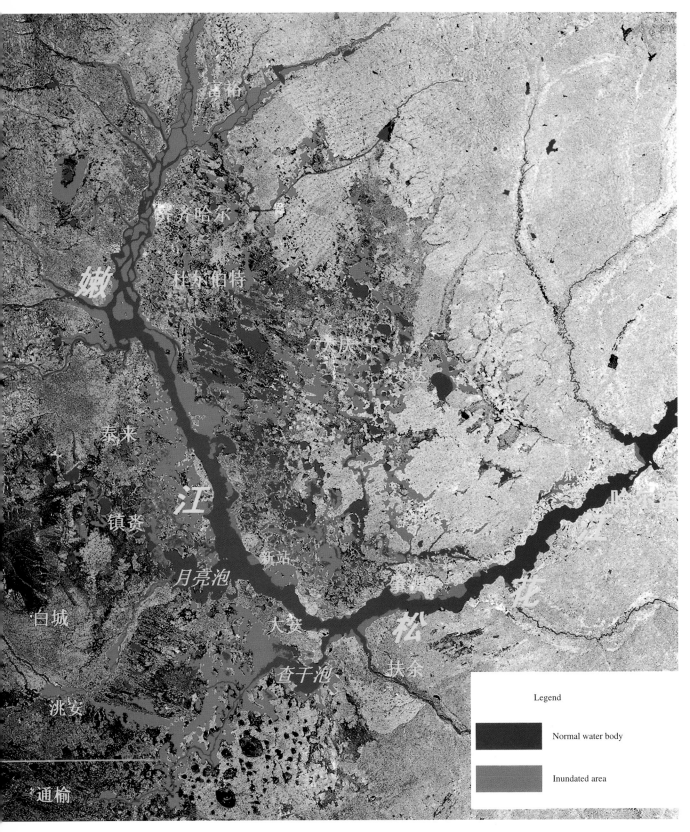

富裕

齐齐哈尔

杜尔伯特

嫩

大庆

安达

泰来

江

镇赉

新站

月亮泡

肇

白城

大安

松

扶余

查干泡

洮安

通榆

Legend

Normal water body

Inundated area

Flood inundated area distribution map of the Nen and Songhua Rivers.

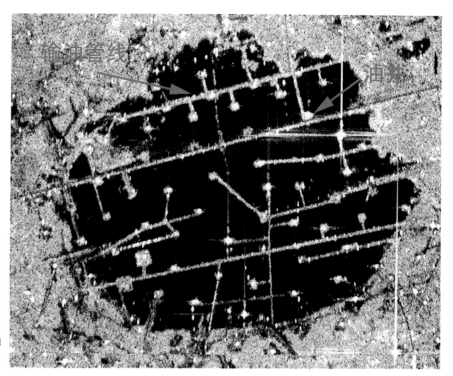

Radarsat image showing the inundated oil wells in the Daqing oil field.

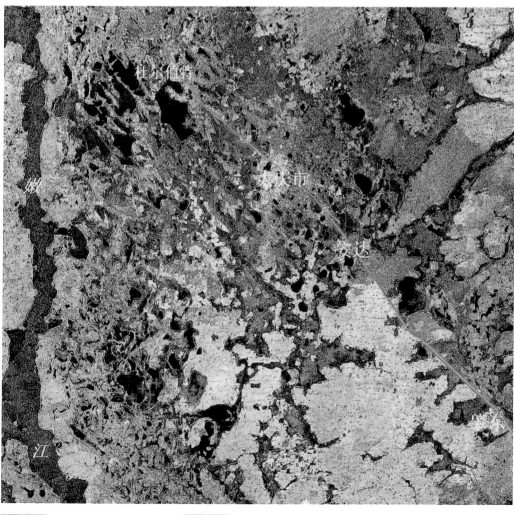

Flood inundated railways, oil wells in Daqing oil field (produced with Radarsat image acquired in August 9, 1998).

 Boundary of inundated areas   Inundated oil wells   Inundated railways

GEOLOGICAL HAZARDS

Landslide

Landslides involving bedrock slippage are a major problem on bank slopes in the Three Gorge area of the Yangtze River. The number of landslides found in this area is 177. The total volume of these landslides is about 1420 million cubic meters. Among these landslides, 33 are considered large-scale (with the volume > 10 million cubic meters). Most of the landslides in this area are bedrock landslides. They occur in strata of clastic rocks with weak intercalations. In the Jurassic and Middle and Upper Triassic strata, there are 20 large-scale landslides. In the carbonate strata, only three small-scale landslides were found. In crystalline rocks, landslides are uncommon.

In this area, landslides in surficial deposits are very rare. Rather, they are comprised of partly or completely revived old landslides in bedrock. The revival mechanisms are mainly rainstorm induction and loading induction (such as the Xintan landslide).

Focusing on the tonal variation of landslides and their contact relationship with surrounding rocks, several landslides were recognized from Radarsat images. From a false color composite multi-temporal Radarsat image

of the Three Gorge area, four landslides (Xintan, Huanglashi, Fanjiaping, Liulaiguan) and one deformation body (Liziya) were identified, of which the Xintan landslide is the largest.

The Xintan landslide is a typical case history of a landslide revival triggered by rear loading. According to historical records, this landslide has slipped repeatedly several times. In the past, navigation along the river has been held up and ships have been turned over and sunk by its slides. The latest large-scale slide occurred on June 12 1985. A great volume of rock blocks and soils, amounting to 30 million cubic meters, slid violently. The town of Xintan was completely destroyed in one fell swoop. Navigation along the Yangtze River was hindered for quite a while. Owing to accurate forecasting, all the inhabitants evacuated from the town in sufficient time. No one was killed or wounded.

False color infrared aerial photograph of the landslide area before sliding.

False color infrared aerial photograph of the landslide area after sliding.

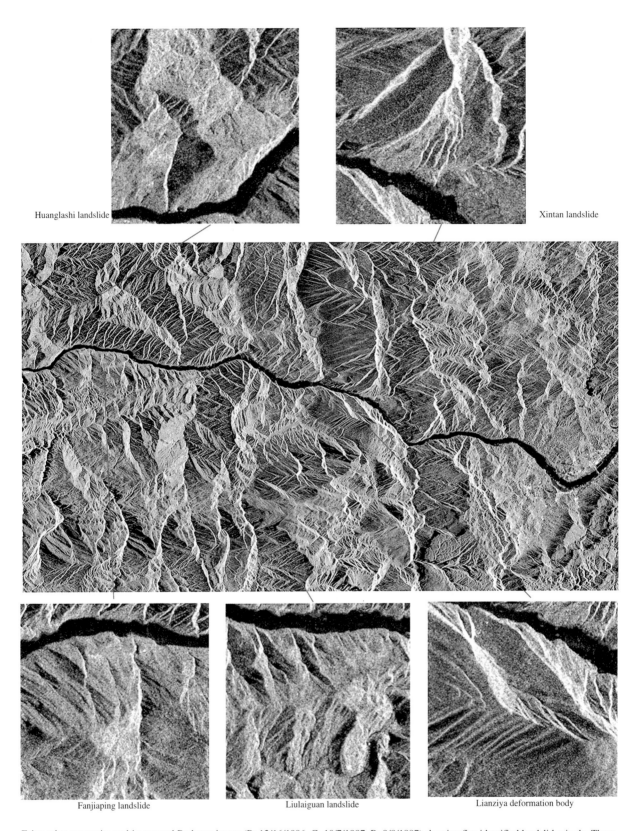

Huanglashi landslide

Xintan landslide

Fanjiaping landslide

Liulaiguan landslide

Lianziya deformation body

False color composite multi-temporal Radarsat image (R: 12/16/1996, G: 10/7/1997, B: 9/8/1997) showing five identified landslides in the Three Gorge area of the Yangtze River.

Earthquake

This false color SIR-C composite, composed of L-HH(R), L-HV(G) and C-HV (B), shows a portion of Mt. He Lan Shan in Ningxia Province. The central coordinates of the scene are 39°04′ 08″ N and 106°15′ E. The incident angle at the image center is 24°43′ 12″. West of the Mt. He Lan Shan is the Jilantai to Linhe depression zone, and to the east is the Yin Chuan faulted basin. Huge concealed faults are developed on both side of Mt. He Lan Shan. The Yin Chuan faulted basin is a Cenozoic basin, controlled by faults oriented in NW and NE directions. Several earthquakes have occurred along a belt including Wuzhong, Pingluo, and Yin Chuan. This area, covered by Holocene strata, is still a depression. A severe earthquake with magnitude of scale 8 occurred to the south of Pingluo City in 1739. The earthquake was controlled by several concealed large faults shown on the geological interpretation map generated from the SIR-C image.

Legend
1. Quaternary
2. Lower Tertiary to upper Pleistocene
3. Archean to Mesozic
4. Fault
5. Epicenter
6. Circular structure

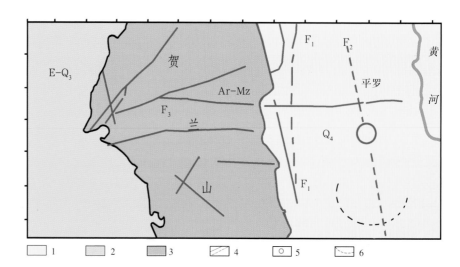

Geological interpretation map of SIR-C image, Pingluo area of Ningxia Province.   0   10km

False color SIR-C image of the Pingluo area, Ningxia Province (R: L-HH, G: L-HV, B: C-HV).   0   20km

GLOBAL CHANGE

Global change studies deal with natural and human-induced changes in the Earth's system from a global perspective. Components of the Earth's geosphere, hydrosphere, atmosphere, and biosphere are, or potentially are, global variables. Numerous activities are required to study a global variable. These activities range from data acquisition to the generation of a data-analysis product, including estimates of the uncertainties in that product. A global measurement often will consist of a combination of observations from a spacecraft instrument and measurements *in situ*.

Scatterometers aboard spacecraft are very useful for providing global coverage data of the normalized radar cross section or radar backsactter coefficient. Thus, scatterometer data is particularly suitable for global change studies. The data has the following distinct characteristics and advantages. (1) Large area coverage and high repeat rate. For example, the ERS-1 scatterometer has an illuminated swath of 500 km and three repeat cycles of 3, 35 and 168 days. (2) Low sampling rate of 25 km $\times$ 25 km for σ facilitates management of a global database.

GLOBAL LAND SURFACE CHANGES

Backscatter Coefficients of Typical Terrain

Global backscattering coefficient images were produced by ERS-1 wind scatterometer (WSC) data acquired at 45° incident angle from forebeam antenna. Sampling windows for statistically analyzing σ° values were located on typical terrain types identified by Kennett and Li (1989). The statistical results were obtained by averaging σ° values of different months from sampling windows of different terrain features. The resultant table suggests that there is a relationship between σ° values and land-cover types. The average σ° value of tropical rain forest ranges from -8 to -7.8dB,

Radar backscatter coefficients of typical terrain features

Terrain features	σ° values (dB)
Tropical rain forest	-8.0 ~ -7.8
Monsoon forest, savanna and crops	-10.8 ~ -9.3
Deciduous forest and needle-leaf forest	-11.5 ~ -11
Temperate grassland and tundra	-15.4 ~ -12.7
Desert	-23.3 ~ -18.5

indicating little changes with seasonal variation. Deserts have the lowest σ° value of global land cover types, ranging from -23.3 dB to -18.5 dB. No matter how the σ° value of different terrain features varies with seasonal changes, a common character is that the distributive extents of land cover types are varied globally in different seasons.

Seasonal Variation in Global Continents

The study for σ° variation of different seasons in the continents of the world used ERS-1 WSC data of March 1996, July 1995, October 1995, and December 1995 to represent spring, summer, autumn, and winter respectively. From a σ° distributive map of these four seasons, a large σ° seasonal variation of different terrain features in the world can be discerned.

Broad-leaf Vegetation

This group comprises mainly tropical rain forests and evergreen broadleaf and deciduous broadleaf forests in the temperate and cold-temperate zones. The forests appear white on ERS-1 WSC images, similar to σ° in the Arctic, Antarctic and Himalayan regions. In comparing four seasonal ERS-1 WSC images, the white color distributions in temperate and cold-temperate zones vary with season, most of them occurring in summer. This indicates the predominance of deciduous broadleaf forest. However, the Amazon and the Congo Basin, which consist mainly of tropical rain forest, hardly

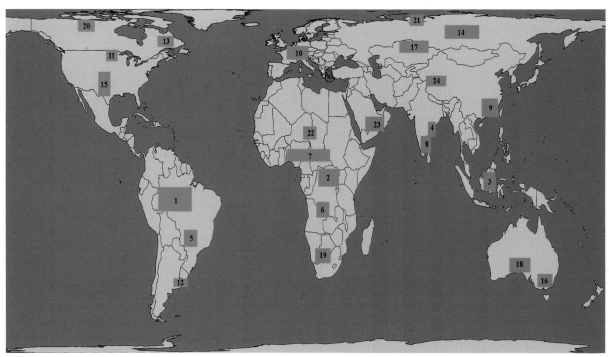

Sampling sites for σ° statistics of typical terrain features in the world.

change with seasons. The fluctuating range of tropical rain forest σ° curves is only 0.2 dB.

Monsoon Forest, Savannah, and Crops

The annual average σ° values of these types range from -9.3 dB to -10.8 dB. Their σ curves indicate that they have conspicuous characters, which change with seasons. For example, for the Monsoon forest in India, σ° values increase from May and become steady from July to October. They begin to decrease from November and drop to their lowest point from December to April of the following year. For the Savannah in Central Africa, σ° values drop from April to August and maintain a steady rise from September to March of the following year. For crops in China, σ° values increase steadily from March to August, and then fall gradually from September to November, reaching their lowest point between December and February of the following year. This indicates that σ° values change with seasons, particularly in southern China.

Deciduous Forest and Coniferous Forest

These types of vegetation are mainly located in the cold-temperate zone, stretching across Eastern Europe and the northern part of North America. The σ° curves reveal some differences between the deciduous forest and the coniferous forest. The σ° values of deciduous forest increase steadily from March to October and then drop from November to February of the following year. From June to September, σ° values of the coniferous forest are the same as those of the deciduous forest, but from October to May of next year they become lower.

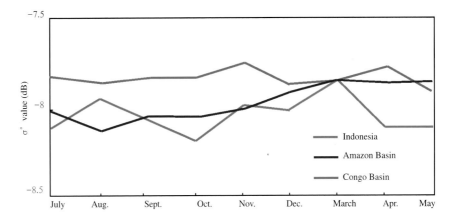

Annual variation curves of σ° values for tropical rain forest.

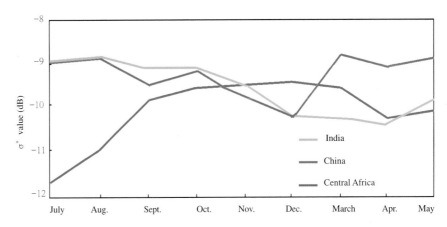

Annual variation curves of σ° values for Monsoon forest, Savannah and crops.

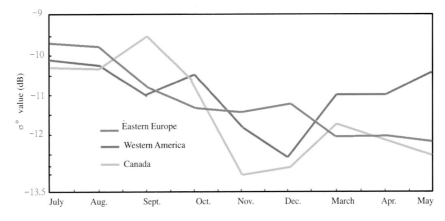

Annual variation curves of σ° values for deciduous and coniferous forests.

215

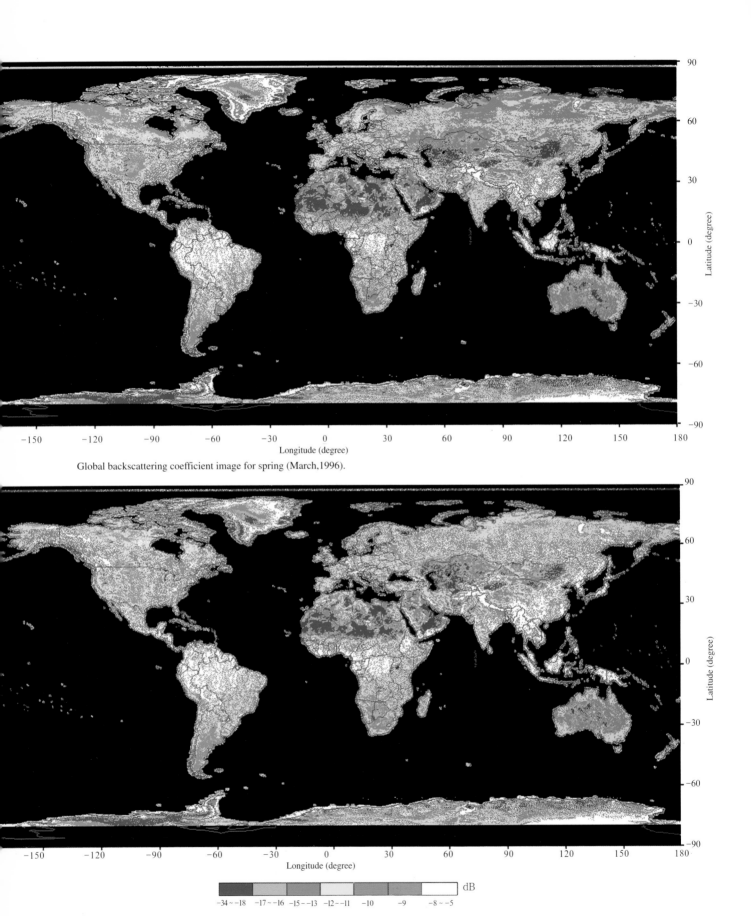

Global backscattering coefficient image for spring (March,1996).

Global backscattering coefficient image for summer (July,1995).

216

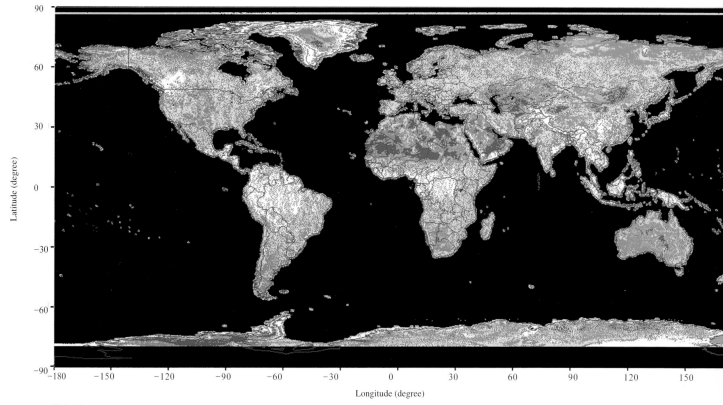

Global backscattering coefficient image for autumn (October,1995).

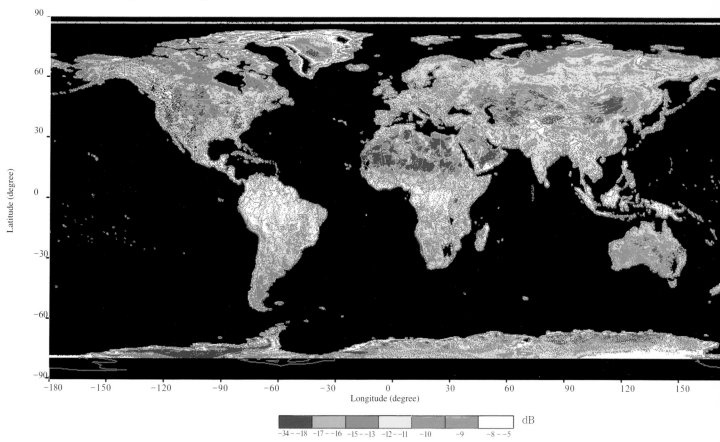

dB
−34 − −18 −17 − −16 −15 − −13 −12 − −11 −10 −9 −8 − −5

Global backscattering coefficient image for winter (December,1995).

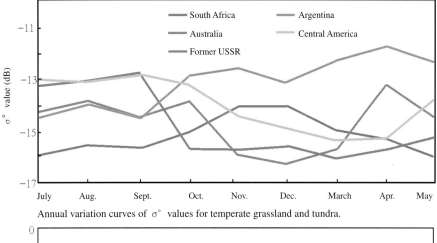

Annual variation curves of σ° values for temperate grassland and tundra.

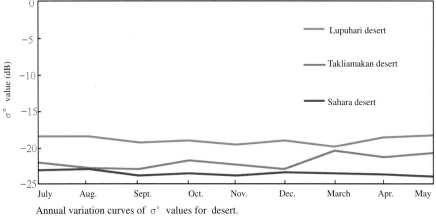

Annual variation curves of σ° values for desert.

The annual variation curves of σ° values for three famous deserts in the world, the Sahara, Lupuhari and Takliamakan Deserts, show that the variation of σ° for each desert is very small. However, the annual fluctuation of σ° values among different deserts is very high and seems to have a relationship to geographic location.

LAND SURFACE CHANGES IN CHINA

Backscatter coefficients of typical terrain

A total of sixteen sites from China were selected for statistical studies to determine the σ° values of different land cover types. The statistical results show that there is a large variation in range for different types of terrain features. The difference between the minimum and the maximum backscattering coefficient is 13.6dB. This is used to identify the different types of land surface coverage. The resultant table shows the corresponding relationship between the type of land covers and σ°.

Temperate grassland zones and tundra are mainly located in the central parts of Europe and Asia, North America and the South part of South Africa. The annual average of σ° values for these type of terrain features range from -15.4 dB to -12.7 dB. They are in light green on ERS-1 WSC images. The annual variation curve of σ° values suggests major changes occur mainly in October and May for the northern hemisphere and southern hemisphere respectively.

The corresponding relationship between land cover types and σ° in China

σ° (dB)	Terrain	Sample Locations
-10.93- -8.56	forest and crops	Loess plateau (8), Da Hingganmt (11), North China plain (13), Chengdu Basin(14), Jianghan plain(15), Jiangnan Hills(16),
-13.69 - -12.72	grassland	Qaidam Basin(3), Tibet plateau(4,5). Northeast China plain(12)
-18.00 - -15.85	steppe desert	Junggar Basin (1), Ordos plateau(9), Inner Mongolia(10)
-21.17 - -19.76	desert	Tarim Basin (2), Badain Jaran Desert(6), Tengger Desert(7)

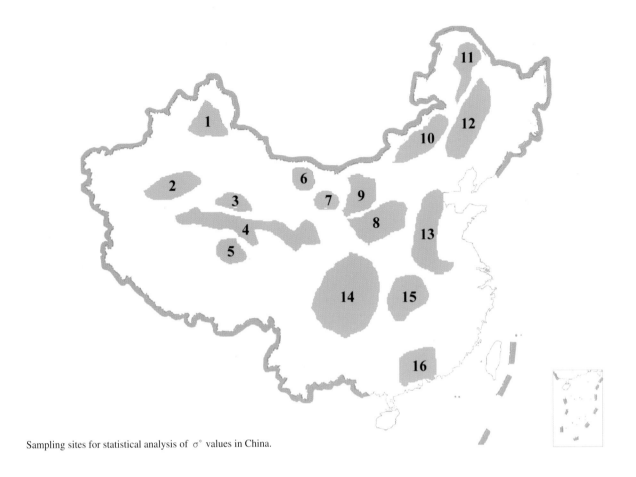

Sampling sites for statistical analysis of σ° values in China.

Seasonal Variation of Land Covers

The ERS-1 WSC images for China over different seasons are enlarged portions of those for the world. Results show that σ° images of China reveal not only the macroscopic landform features, but also the strong seasonal variations of these terrains.

Forest and Crop

The annual average value of σ° for this type of terrain ranges from - 10.93dB to - 8.56dB. On the ERS-1 WSC images, the distribution area for this type of terrain is represented in red color, and it is easily seen that this type of terrain varies with seasons. The annual variation curves of σ° values for this type of terrain show some marked differences in σ° values in the South and North of China. In southern China, annual variation of σ° for this type of terrain is relatively small and its fluctuation range is only 2 dB. The σ° values are lower from December to February of following year. In Northern China, the annual variation range for this type of terrain is much higher and its fluctuation range can reach 4 dB. The σ° values are lower from October to March of following year .

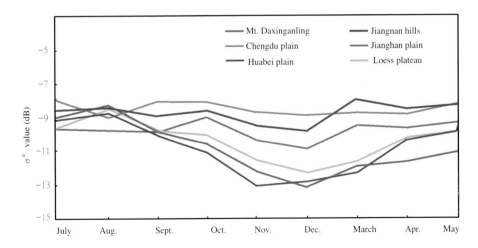

Annual variation curves of σ° values for forests and crops in China.

219

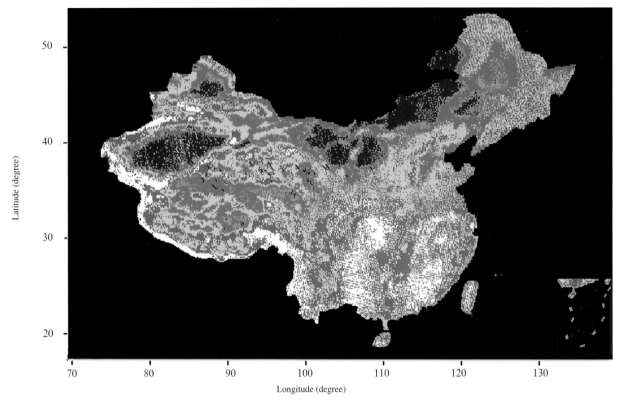

Backscattering coefficient image for spring in China (March,1996).

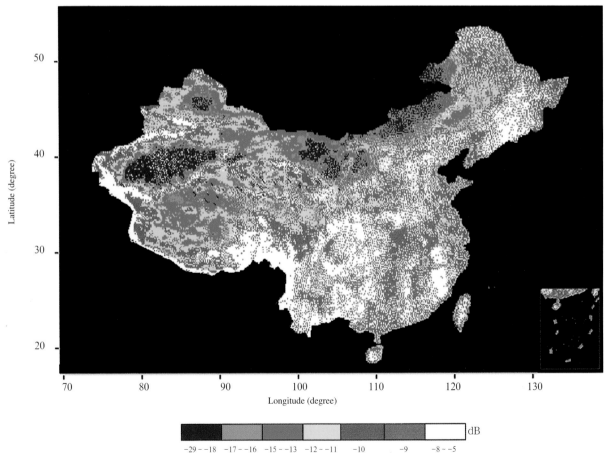

dB

-29 ~ -18 -17 ~ -16 -15 ~ -13 -12 ~ -11 -10 -9 -8 ~ -5

Backscattering coefficient image for summer in China (July,1995).

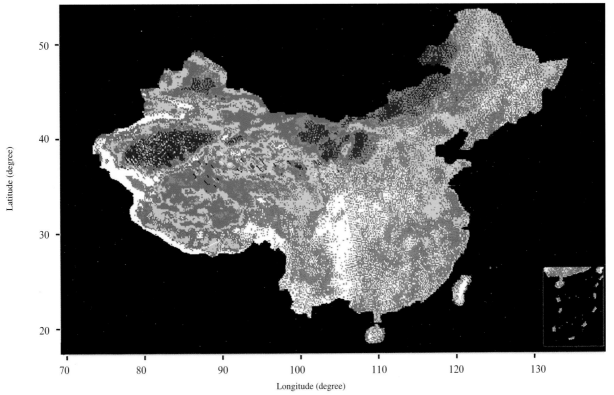

Backscattering coefficient image for autumn in China (October,1995).

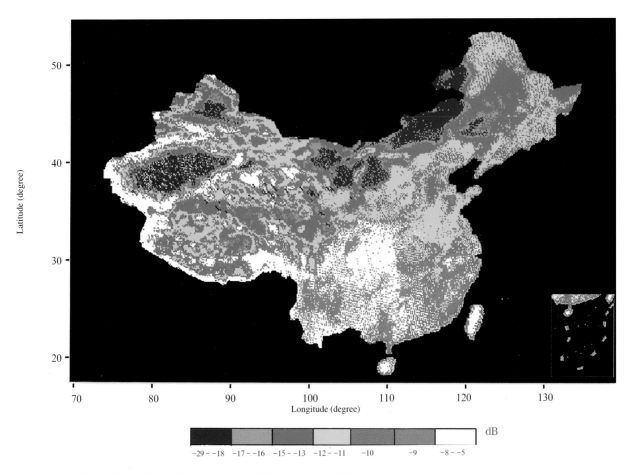

Backscattering coefficient image for winter in China (December,1995).

Grassland

Grassland here refers to terrain features having similar $\sigma\degree$ values to those determined for grassland for global statistics. In China, Grassland is mainly distributed in the North China Plain, the Qaidam Basin, and the Tibet Plateau. The annual average value of $\sigma\degree$ for this type of terrain ranges from -13.69 dB to -12.72 dB .

Steppe Desert

For steppe desert, annual average $\sigma\degree$ values range from -18.00 dB to -15.85dB. Such values are mainly distributed in the Junggar Basin and the Ordos and Inner Mongolia Plateaus. The annual variation curves of these areas show that the variation of $\sigma\degree$ values for the Junggar Basin is relatively steady, with annual fluctuation of only 2 dB. The annual variation of $\sigma\degree$ values for the Ordos and Inner Mongolia plateaus are much higher, with fluctuations in the order of 3dB for Ordos Plateau and 4dB for Inner Mongolia.

Desert

The values of $\sigma\degree$ over deserts range from -21.17 dB to -19.76 dB. Such values are found in Takliamakan, Badain Jaran, and Tengger Deserts. The annual variation curves of $\sigma\degree$ values for these deserts reflect that the distribution of $\sigma\degree$ for the Takliamakan Desert is extraordinarily steady all year round, except in April; $\sigma\degree$ variations for the Badain Jaran and Tengger Deserts are basically the same.

Among these deserts, the Badain Jaran Desert is the driest in China.

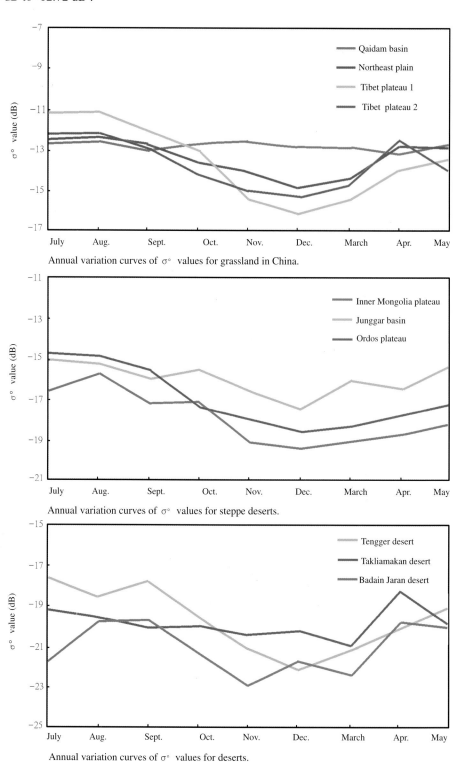

Annual variation curves of $\sigma\degree$ values for grassland in China.

Annual variation curves of $\sigma\degree$ values for steppe deserts.

Annual variation curves of $\sigma\degree$ values for deserts.

RADAR SCATTERING CHARACTERISTICS AND INFORMATION EXTRACTION

Understanding radar scattering characteristics of terrain features is an essential basis for analyzing radar imagery. This section analyses radar scattering characteristics of crop, forest, lithology and other typical terrain features by means of comparing the effects of different wavelength, polarization and temporal SAR data. This could theoretically enhance the understanding of feature recognition mechanisms referred to in the previous sections.

Information extraction from a radar image is a key technique for recognizing terrain features.

This section presents results of using neural network and fractal analysis for extracting information and also introduces some results from analyzing polarimetric SAR and interferometric SAR data.

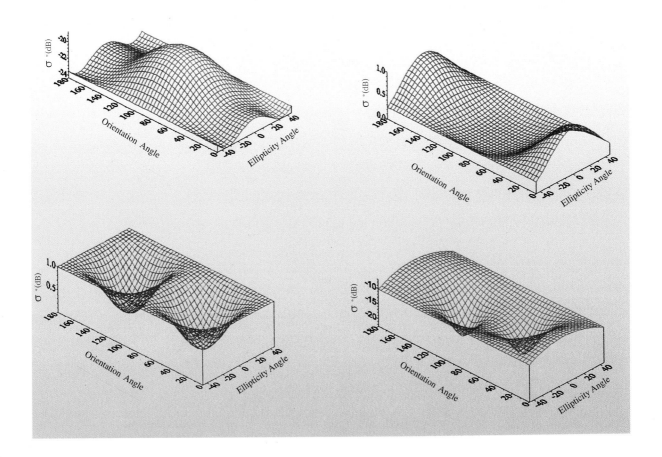

GROUND RADAR BACKSCATTER

Crops

Effect of Radar System Parameters on Targets Backscatter

The backscatter coefficient recorded on synthetic aperture radar represents characteristics of targets such as dielectric constant, surface roughness, geometric shape, and orientation. It also depends on radar system parameters such as wavelength, polarization, incident angle, and spatial resolution. These radar system parameters have significant impact on the backscatter behavior of targets.

A 3-D visualized figure of backscatter intensity was produced from the radar image taken by the Canadian CV-580 airborne radar system. There are four water ponds separated by small dirt roads; three of the ponds are utilized for fish farming and one for growing Euryale ferox, which is an aquatic plant in the same family as lotus. In the figures labeled C-HH and X-HH, returns show there is no difference among the four ponds. However, in figures C-VV and X-VV, returns show there is strong contrast between the fishponds and the *Euryale ferox* pond; however, there is little difference in the C-HV backscatter. *Euryale ferox*, with large flat leaves laying on the surface of water, has a higher VV backscatter than the HH polarization. The flat leaves act as a slightly rough surface with a horizontal structure compared to the pure water surface acting as a specular reflector. Generally speaking, a slightly rough surface has higher VV polarization backscatter than HH polarization backscatter.

The comparison figure shows the variability of backscatter intensity for various targets. Rice fields and harvested rice fields vary dramatically with polarization, showing a wide range of backscatter contrast at C-band HH and VV polarizations and lesser variation at X-band HH and VV polarizations because the harvested rice

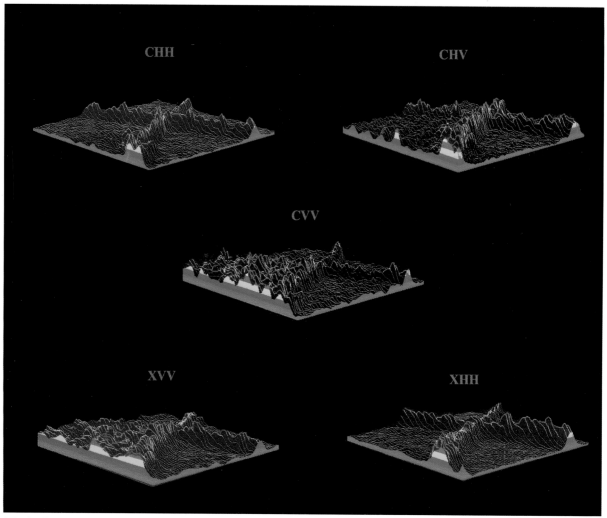

3-D visualized backscatter intensity comparison between water and *Euryale ferox* varying as a function of wavelength and polarization.

fields contain water-saturated soil and rice stubble. The water saturated soil has a much higher dielectric constant than the mature rice, which produces a strong radar return. For sugarcane, banana, or cassava, the HH polarization backscatter intensities are slightly higher than for the VV polarization, which confirms a higher attenuation at the VV polarization for vegetation with a vertical structure.

This figure shows the intensity comparison of rice and water as it varies with wavelength and polarization. It was produced by averaging the digital number (DN) for pixels taken from training areas of the SIR-C/X-SAR image with L-band, HH and HV polarization; C-band, HH and HV polarization; and X-band, VV polarization. As expected, rice exhibited a much stronger response than water at shorter wavelengths. On April 18th, the rice was in a recovery stage just after transplanting and was starting the seedling development period. The height of the rice seedlings was about 20-30 cm. However, even though the rice only had scattered seedlings standing in open water (without a closed canopy), the long wavelength (L-band) radar exhibited the difference between rice and water. Rice seedlings produce a double bounce effect between the water surface and rice stem. If we look at the decibel values (dB) taken from the sampling areas of the same image, it is noted that the backscatter coefficients of rice are 13 dB higher than water at C-band, and 9 dB higher than water at L-band.

It was also found that in the early growth stage rice has higher DN at HH polarization than at HV polarization. The difference in HH

polarization and HV polarization was found to be 8 dB for L-band but only 4 dB for C-band, implying a presence of volume scattering for C-band.

The radar cross section is strongly related to the wavelength. In the case of rice, the radar cross section in C-band is larger than L-band. For HH polarization, it is 2 dB larger than L-band, and for HV polarization, it is 6 dB larger than L-band. Rice is a vertically oriented plant, so the vertical polarization attenuation is higher than the horizontal polarization. The backscatter of the target in H polarization is therefore higher than V polarization.

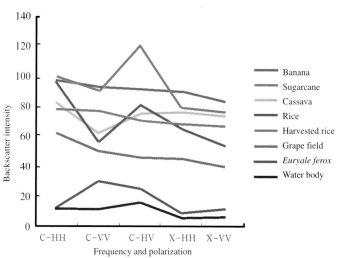

Comparison of backscatter intensity of crops as a function of frequency and polarization.

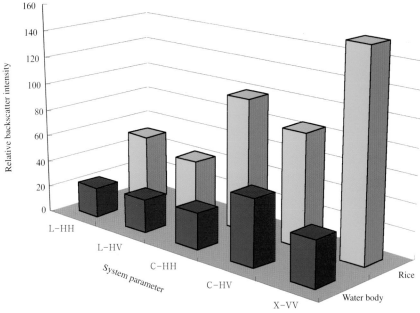

Comparison of intensity of rice and water as a function of wavelength and polarization.

Backscatter Behavior of Crops

This study is aimed at understanding the backscatter behavior of vegetation or crops during their crop calendar and the relationship between the vegetation and crop structure parameters to their backscatter coefficients. The first figure (line graph) shows the backscatter coefficients produced from calibrated RADARSAT image for the autumn of 1996. The second figure (block diagram) is the relative backscatter intensity produced from uncalibrated RADARSAT imagery acquired from spring to autumn of 1996. The third figure (another line graph) is the backscatter coefficients produced from calibrated RADARSAT imagery acquired in spring of 1997. The backscatter from bananas remains almost constant as a function of time. Over time, the structure of the banana plant does not change, but the water content of the banana plants increases in the wet season and decreases in the dry season. The backscatter of *Euryale ferox* changes over time. From spring to summer, the leaves of *Euryale ferox* grow gradually. The backscatter increases as the leaf index rises. The leaf area growth reaches its peak in August. Afterwards, as the leaves dry out with further maturity, the backscatter decreases. The backscatter intensity of lotus is an interesting curve. Its backscatter intensity increases gradually from April as the leaf index of lotus increases, then drops as the lotus matures until harvest in November. The end of July is harvest time for spring rice. The backscatter intensity curve of mature spring rice

has a break in August due to the harvest. The beginning of August is the transplanting time for autumn rice. Sugarcane reaches its peak growth in May or June and generates the highest backscatter in its growth cycle. The backscatter coefficients of forest and grassland show little variation over time. Therefore, their DN values change little over time. During the summer, forests and grasslands have a slightly higher backscatter than in autumn because their water content is normally higher in summer.

Backscatter coefficients of *Euryale ferox* are relatively low; the highest value is less than -10 dB. The backscatter coefficients of buildings produce the highest value, larger than any other targets, normally greater than -25dB. The backscatter coefficient of a building is not only related to cross section, but primarily to orientation. Calm water consistently has backscatter coefficients lower than -20 dB.

This study is aimed at understanding the backscatter behavior of rice during its growth cycle and the relationship between rice structure parameters and their

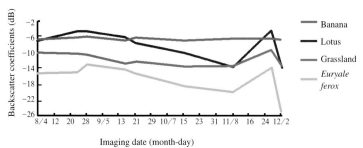

Backscatter coefficients of vegetation as a function of time (autumn of 1996).

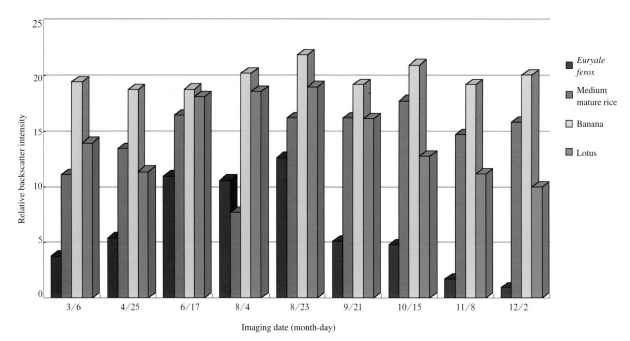

Relative backscatter intensity of vegetation as a function of time (spring, summer, and autumn of 1996).

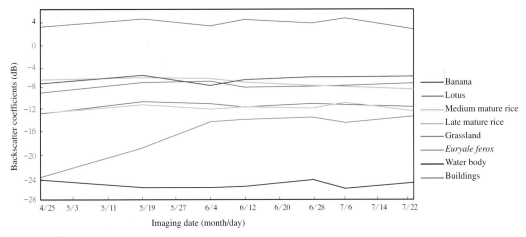

Backscatter coefficients of targets as a function of time (spring of 1997).

Field Photos of main crops at Zhaoqing test site (these are banana, vegetable lands, *Euryale ferox* and lotus)

Field photos of main crops at Zhaoqing test site (these are younger and mature sugercanes, grasslands and vegetable lands)

backscatter coefficients. In the figure produced from calibrated RADARSAT images acquired in autumn of 1996, there are five types of rice: 1) (RM) medium mature rice, 2) (RLT) late mature rice, 3) (RLTP) rice transplanted about 25 days later than the normal spring rice due to a cold spring and low temperatures in 1996, 4) (RSP) spring rice, and 5) (RAT) autumn rice. The backscatter coefficients of late mature rice are slightly higher than medium mature rice. As a higher yield rice species, the late mature rice has slightly higher backscatter coefficients than medium mature rice. RLTP life span is relatively short, and the backscatter coefficients are lower. The backscatter coefficients of RM, RLT, and RLTP produced from the RADARSAT images taken after April 25th increase gradually until the rice is harvested. RSP and RAT rice types demonstrate how farming activities can change during the year, altering from fish ponds to rice fields. This practice is very common in the Zhaoqing area as well as many other parts of China. Switching from fish pond to rice generally implies that the soil quality of the field is less than excellent and that the rice yield is lower than average.

In the figure produced from the DN number for four types of rice from uncalibrated RADARSAT images acquired during the growing season of 1996 there are four rice designations: 1) R2 represents medium mature rice, 2) R3 represents late mature rice, and 3) R4 and R1 represent single rice crops in spring and autumn respectively.

In the figure produced from calibrated RADARSAT images acquired in the spring of 1997, there are four major types of rice, each with a different growth cycle. Early mature rice is represented as R4, medium mature rice as R1, medium-late mature rice as R3, and late mature rice as R2.

When RADARSAT acquired data over the

Zhaoqing test site, ground measurements of rice structure parameters and water content were taken on a real-time or near real-time basis from spring to summer of 1997. The first figure shows the water weight content of rice. Rice has high water weight content in its seedling development period and ear differentiation period. At the end of the ear differentiation period, the water weight

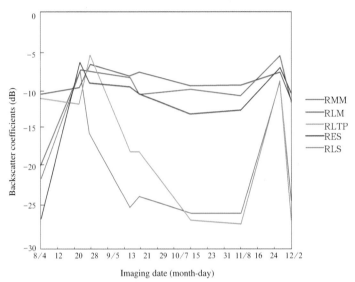

Backscatter coefficients of rice as a function of time (autumn of 1996).

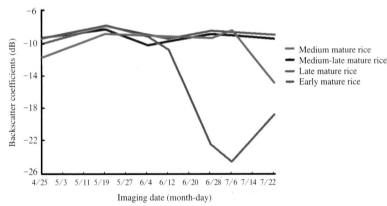

Backscatter coefficients of rice as a function of time (spring of 1997).

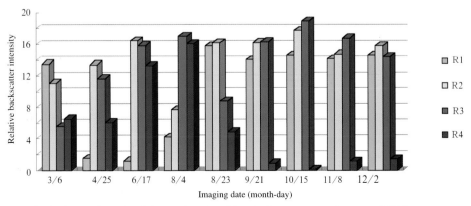

Relative backscatter intensity of rice as a function of time (spring, summer, and autumn of 1996).

content decreases slowly, reaching its lowest point during the mature period. It was found that the water weight content of rice at its early stage is about 20-30% higher than during the maturing period. The third figure shows the changes of rice height and leaf length. Measurements were taken for the height of the rice stem and the length of leaf 1, leaf 2, leaf 3, leaf 4 and leaf 5 from bottom to top. Starting from the transplanting period, the height of the rice stem increased from 20 cm to about 100 cm and reached its peak height during the heading period. It remains at that height for a period, then drops by about 5 cm, before stabilizing again at that height. The second figure illustrates the simulation results for backscatter coefficients from the multi-temporal RADARSAT image.

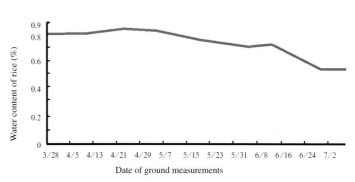

Water content of rice as a function of time (spring of 1997).

Simulated backscatter coefficients of rice.

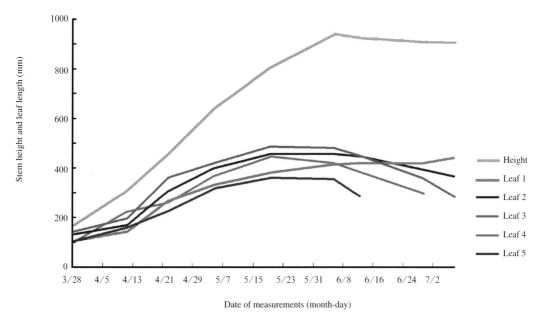

Date of measurements (month-day)

Structure parameters (height and length of leaves) of rice as a function of time (spring of 1997).

How resolution affects target recognition

The definition of the resolution for a SAR image differs from an optical sensor image. SAR has both range resolution and azimuth resolution. The range resolution is dependent on the bandwidth of the transmitted signal. The azimuth resolution is dependent on the bandwidth of Doppler frequency. Pixel spacing of an SAR image is not equal to the spatial resolution. Usually, the pixel spacing is half of the resolution. The Pearl River Delta site has been selected to demonstrate the influence of resolution on target recognition. This discussion is meaningful for many applications and data requests.

For a distributed target, resolution only affects its edges, making the edges obscure. In contrast, for a point target or a small target, resolution is very important. As resolution decreases, the smaller targets may disappear. On the left corner of the image, there is a large amount of sugarcane. The sugarcane fields have well-defined boundaries on standard mode RADARSAT image, but the boundaries become more obscure as the resolution decreases. The boats in the river are very distinctive in the high resolution image, but it is difficult to see the small boats in the low resolution image. From north to south, there are a number of bridges over the river. They are clearly visible in the standard mode image, but nearly invisible in the wide scan SAR image. Therefore, when we make a data request, it is extremely important to request the correct balance between data resolution and coverage. The most cost-effective product is a desired goal.

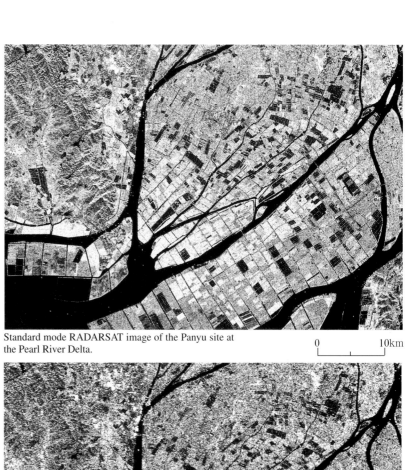

Standard mode RADARSAT image of the Panyu site at the Pearl River Delta.

0   10km

Scan SAR narrow-mode RADARSAT image of the Panyu site at the Pearl River Delta.

0   10km

Scan SAR wide-mode RADARSAT image of the Panyu site at the Pearl River Delta.

0   10km

Forest

Relating Radar Backscatter to Forest Stand Parameters

It is important to study the relationship between radar backscatter and forest parameters. Understanding the relationships between radar backscatter and phenological variables can improve radar backscatter models of tree canopy properties and assist in developing a radar-based scheme for monitoring forest phenological changes.

The airborne GlobeSAR data collected were uncalibrated. We utilized the radar backscatter intensity instead of the radar backscatter coefficient after correcting for antenna pattern and geometric rectification. To compute the mean value of SAR data for a stand of forest, the stand was located on the SAR imagery and the largest possible window within the stand was extracted. For each stand, at least 200 image pixels were averaged. Forest stand parameters of ground measurements consisted of stand basal area, density, mean height, and diameter at breast height (dbh).

It can be seen that there is a high correlation between the radar backscatter intensity and the mean stand height and dbh. As the mean dbh and height of trees in the stands increase, the C-HH, C-VV, X-HH and X-VV backscatter increase. The observed increase in backscatter may be attributed to the increase in tree size. On the other hand, there is almost no relationship between the radar backscatter and the stand density or between the radar backscatter and the stand basal area. Therefore, the stand density and basal area are main factors affecting the variation in the SAR backscatter.

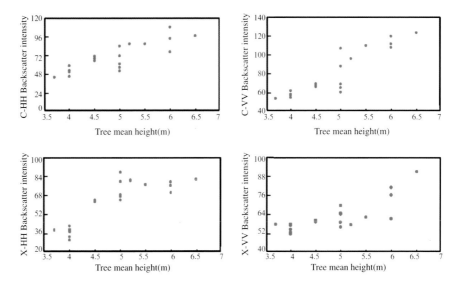

Tree mean height vs. C- and X-band, HH and VV polarizations backscatter intensity for a pine forest in the Zhaoqing test site, Southern China.

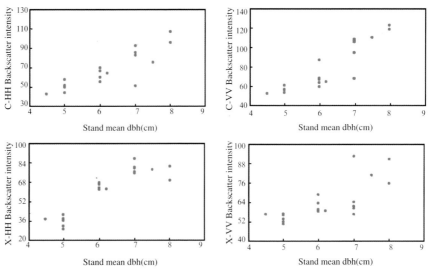

Stand mean dbh vs. C- and X-band, HH and VV polarizations backscatter intensity for a pine forest in the Zhaoqing test site, Southern China.

231

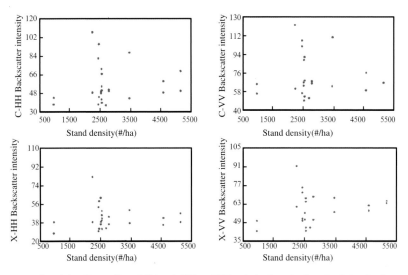

Stand density vs. C- and X-band, HH and VV polarizations backscatter intensity for a pine forest in the Zhaoqing test site, Southern China.

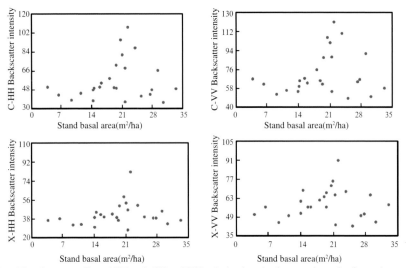

Stand basal area vs. C- and X-band, HH and VV polarizations backscatter intensity for a pine forest in the Zhaoqing test site, Southern China.

Modeling Forest Radar Backscatter

Understanding scattering mechanisms is important for forest microwave remote sensing applications. In this study, a discontinuous canopy model was used to model radar backscatter of a Zhaoqing pine forest, and comparisons were made between SIR-C data and model predictions for the test forest stand.

SIR-C data operating in L- and C-bands, and HH, HV, VH and VV polarizations were acquired in April of 1994. The multifrequency and multipolarization image (R: L-HH, G: L-HV, B: C-HV) shows the pine stands as green. The backscatter coefficients extracted from the SIR-C image and predicted from the model (see table) demonstrated that the model is able to predict the HH and VV backscatter of L- and C-bands. In addition, the

HH and VV backscatter coefficients are higher in L-band than in C-band. The models suggest that longer wavelength radar would enable us to derive information about particular parts of the canopy due to its penetration. On the other hand, the backscatter coefficients of L- and C-bands are also higher in co-polarizations than cross-polarizations, and in HH polarization than in VV polarization. The model showed that the radar backscatter is stronger for co-polarizations than cross-polarizations, and for HH more than for VV.

In the discontinuous canopy backscatter model, we considered four backscattering components, including surface backscatter, crown volume scattering, crown-ground interactions, and double-bounce trunk-ground interactions. It was observed that in C-band, the radar backscatter of the pine stand is almost entirely due to crown volume scattering at all polarizations. At this frequency, the incident microwave energy cannot reach the trunk and ground surface, and hence the radar measurements are not directly sensitive to target factors such as trunk heights and surface parameters. In L-band, the longer wavelength can penetrate the crown layer to a larger extent to show information about the lower branches, trunk, and ground conditions under the forest canopy. Thus the L-band radar backscatter is seen to be primarily due to

L- and C-bands radar backscatter coefficients of pine forest

	Extracted from SIR-C Image		Predicted from the model σ° (dB)($\theta =28^{\circ}$)
	σ°(dB)	sd*	
L-HH	-8.1	1.1	-9.5
L-HV	-16.1	1.0	-20.0
L-VV	-9.0	0.8	-10.9
C-HH	-11.5	1.0	-12.2
C-HV	-14.1	0.9	-17.2
C-VV	-12.9	0.5	-12.6

*sd:Standard deviation

double-bounce trunk-ground and crown-ground interactions, with a small but still significant amount of crown and surface contributions.

At all frequencies, the magnitude of the cross-polarized return is lower than that of the co-polarized return because the cross-polarized backscatter is mainly due to crown volume scattering. At L-HV, the tree crown not only contributes to the backscatter but also results in attenuation when the incident waves penetrate the canopy. Hence the radar backscatter is higher at L-HV than at C-HV.

At C-HH, crown-ground and trunk-ground backscatter show an increase as the incident angle increases. The trunk-ground double bounce scattering also increases with less than 50 degree incident angle. Thus radar backscatter is higher for both HH and VV polarizations than for HV, and for HH than for VV.

At L-HH, the double bounce mechanism is mainly

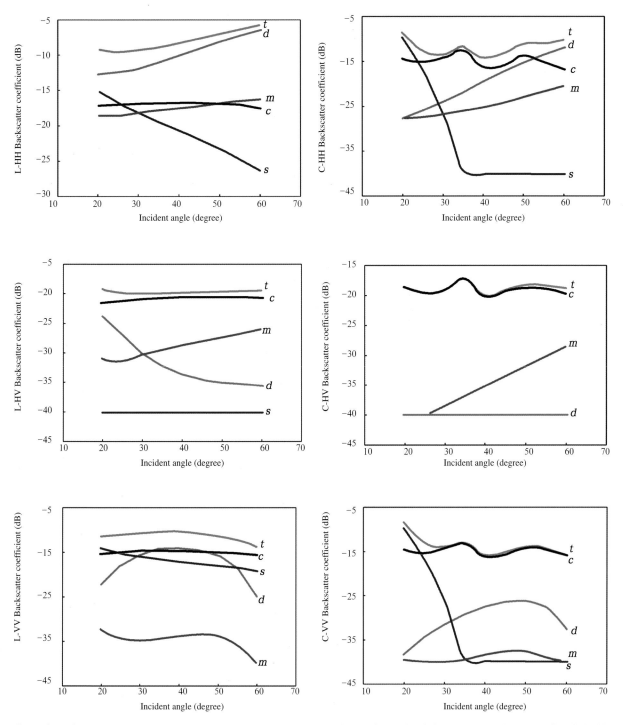

Comparison of model components at L- and C-bands for pine stands at the Zhaoqing test site. (Abbreviations: c=crown volume scattering; d=double-bounce trunk-ground interactions; m=crown-ground interactions; s=surface backscatter; t=total backscatter.)

Multifrequency and multipolarization SIR-C image of the forest area at the Zhaoqing test site, Guangdong Province (R: L-HH, G: L-HV, B: C-HV).

scattering sources and shows an increase as the incident angle increases. However at L-VV, crown volume scattering and trunk-ground double-bounce interactions contribute to radar backscatter. The surface backscatter becomes lower as the incident angle increases, and the double-bounce scattering is the highest over the 35 to 45 degree incident angle range. The crown volume scattering has less variation and demonstrates that radar backscatter is lower at L-VV than at L-HH.

Rocks

Lithological Classification with Polarimetric Data

With the Theorem of Target Decomposition (TD), we use the coherent matrix data in the test site of Qinghe,

Xinjiang and decompose the backscattering intensity of surface rocks into three parts: Single Scattering, Double Scattering, and Cross Scattering. Then, we can calculate the entropy of target scattering.

In the false color image of SIR-C L-band and C-band backscattering coefficients ($\sigma^{\circ}_{LHV} + \sigma^{\circ}_{CHV} + \sigma^{\circ}_{LVV}$), we selected nine classes of targets according to the geological map (Dry River, Alluvial Fields, Sandstone, Plagio-porphyrite, Pebbly-sandstone, Schist, Hard-arkose, Granitite, and Light Granite). Then we extracted 200 samples for each target and obtained six kinds of TD images, LHH-LVV Coherence image, Phase Difference image, and Classification Result image.

The TD images display that Target Decomposition can show the backscattering differences from various types of rocks; L-band shows greater differences than C-band. For example, Plagio-porphyrite has more cross scattering, and Dry River has more double scattering.

For all the targets, Dry River has the lowest LHH-LVV coherent coefficient value and the highest Phase Difference. We can interpret it correctly first.

We use three kinds of data collection and presentation techniques to make lithological classification: (1) $\sigma^{0}_{LHV} + \sigma^{0}_{CHV} + \sigma^{0}_{LVV}$ as the false color image; (2)TD data $+\sigma^{0}_{LHV} + \sigma^{0}_{CHV} + \sigma^{0}_{LVV}$ to display the role of TD data in classification; and (3) TD data $+ \sigma^{0}_{LHV} + \sigma^{0}_{CHV} + \sigma^{0}_{LVV}$ Phase Difference data to compare the data fusion of additional data. After evaluation and comparison, we found that the third method had a high precision and high error. The first method had a low precision and

high error. The second method (or TD method) had the highest precision and lowest error. We concluded that for this experiment TD data is the more effective technique for accomplishing lithological classification.

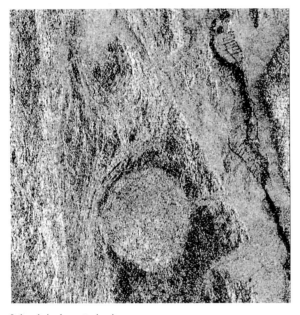

L-band single scattering image.

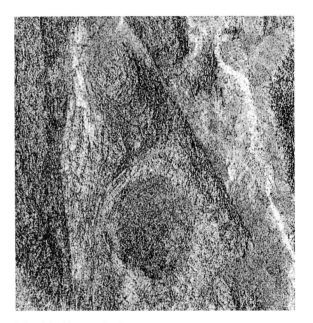

L-band double scattering image.

L-band cross scattering image.

L-band scattering and entropy images of Qinghe test site.

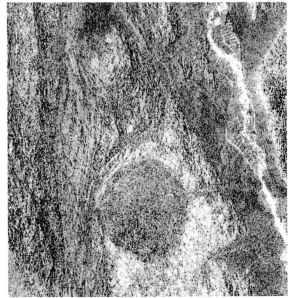

L-band entropy image.

SIR-C backscattering coefficients image ($\sigma^0_{LHV} + \sigma^0_{CHV} + \sigma^0_{LVV}$).

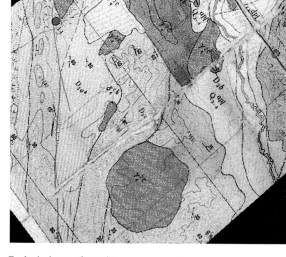

Geological map of test site.

0   5km

1--Dry River   6--Schist
2--Alluvial Fields   7--Hard-arkose
3--Sandstone   8--Biotite Granite
4--Plagio-porphyrite   9--Muscovite Granite
5--Pebbly-sandstone

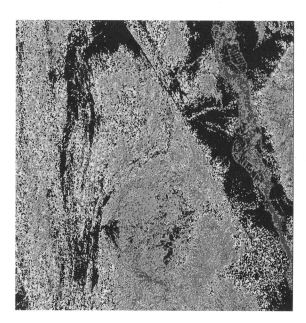

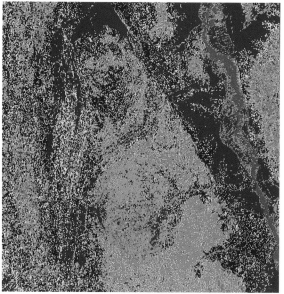

Classification image of three kinds
of methods.
Lower left: First type data
Mid-right: Second type data
Lower right: Third type data

Dry River
Alluvial Fields
Sandstone
Plagio-porphyrite
Pebbly-sandstone
Schist
Hard-arkose
Biotite Granite
Muscovite Granite

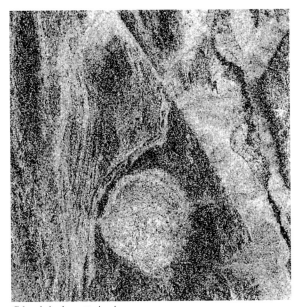

C-band single scattering image.

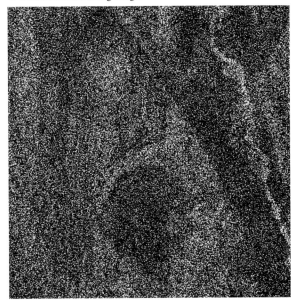

C-band double scattering image.

C-band cross scattering image.
C-band scattering and entropy images.

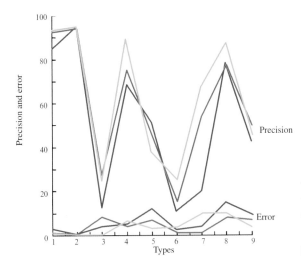

C-band entropy image.

Precision

Error

—— Classification using backscatter coefficients, TD data and phase difference data

—— Classification using backscatter coefficients and TD data

—— Classification using backscatter coefficients

Precision and error comparison of three kinds of methods

Extraction of Bare Surface

There are two alluvial fans in the SIR-C L-band false color image ($\sigma^{0}_{HH} + \sigma^{0}_{HV} + \sigma^{0}_{VV}$), with F1 in the lower left corner and F2 in the upper right corner. For each alluvial fan, we made three profiles along the inner, middle, and edge, and selected 50 samples to analyze the differences of dielectric constant and roughness of the two fans. Finally, we collected the dielectric constant image and rms height image. The first image displayed a range in dielectric constant from 2.0 to 4.28, and the second image displayed a range in surface roughness from 0.0 cm to 2.9 cm.

The dielectric constant range for both alluvial fans ranged between 2.0 to 4.0. But the deviation curve of F1 was smoother than F2, illustrating that the surface physical characteristics are spatially more variable.

Roughness can be displayed in rms height and coherent length. For F1, both parameters decrease slowly from the inner

to the edge of the fan, and the deviation value for them is relatively low. But for F2, both the rms height and coherent length have a higher deviation value than F1 and change in a more random fashion, although F2 shows the same decreasing trend from the inner to the edge as F1.

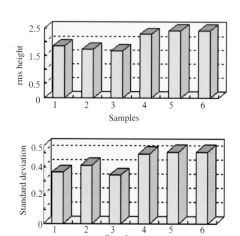

The rms height and deviation of two fans.

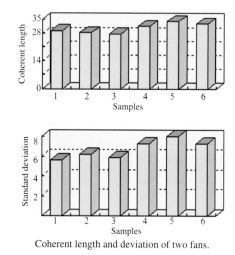

Coherent length and deviation of two fans.

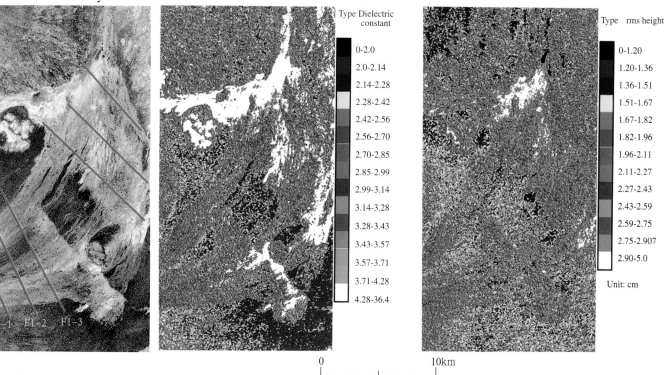

Extraction of bare surface parameters of L-band SIR-C polarmetric data (Left: backscatter coefficients; Middle:dielectic constant; Right: rms height)

INFORMATION EXTRACTION

Information Extraction from SAR Image Using Fractal Method

A fractal is defined as a measurement of irregularity and fragmentation of complex targets. It contains considerable geometrical information and reveals the self-similarity and proportion of a fractal body. It is been proven that the fractal characteristics of a 2-D gray-level image can be used to reveal the fractal features of a 3-D natural surface. Image processing with a small sliding window, we get a fractal character map by calculating the value of fractal dimension of the central pixel of the windows according to the superficial area-volume relation method. Since the superficial area of a fractal image is considered to be a measure of the complexity of the gray surface, the fractal dimension is related to the change in the distribution in complexity of the image surface and has been effectively used on radar images with significant noise and speckle. The above image is the raw data of GlobeSAR of the Zhaoqing test site. Pixel size is 6.25m × 6.25m. Below the image is the D character map extracted using the fractal method. The edges of targets are detected clearly.

GlobeSAR image of the Zhaoqing test site.

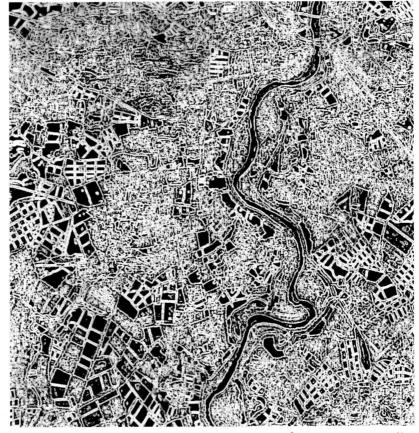

D-character map extracted by the fractal method.

0   3km

239

Learning Vector Quantization Neural Network for Land Cover Classification

The principle of vector quantization is based on using discrete determined vectors to approach continuous stochastic vectors based on the minimization of the error cost function. This means that the determined vectors can represent the distribution of the original continuous vector set with the least error. Kohonen (1988) proposed utilizing the learning vector quantization method in combination with neural network. A learning vector quantization neural network is self-adaptive, has learning ability, and has been successfully used for pattern recognition. A learning vector quantization neural network within a modified learning method can have both the ability of feature extraction and classification. When used for land cover classification in the Zhaoqing area and texture was considered, very high accuracies were obtained and even roads could be classified correctly.

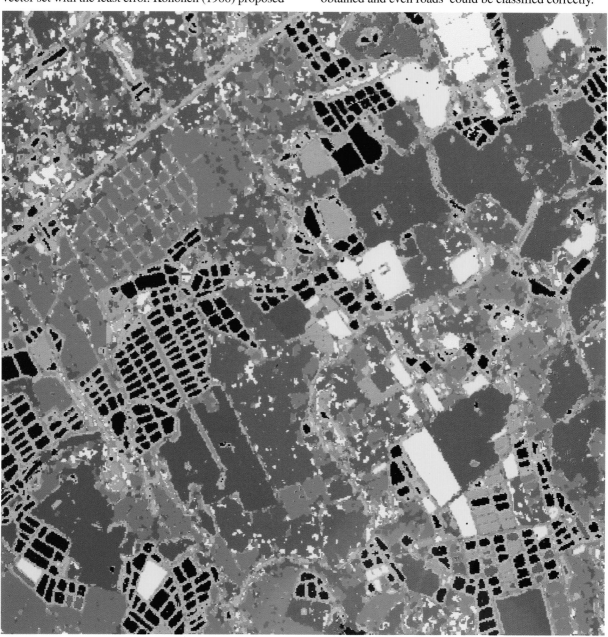

Five classes of paddy fields, each with different transplanting period

Mid-maturing rice | Late-maturing rice | Fishing pond/ Late-maturing rice | Fishing pond/ Early-maturing rice | Early-maturing rice, transplanted ~20 days later

Water body | *Euryale ferox* | Building | Road | Road and tree

Interferometric SAR Data Processing

By acquiring complex SAR images of the same area from repeated over-flights or single-pass flights with two antennas, a 3-D DEM (Digital Elevation Model) can be generated from INSAR data. Here we used SIR-C data of the Karakax Valley area in the Western Kunlun Mountains of China to illustrate data processing procedures.

First, from INSAR data, two single look complex images of amplitude and phase can be separated. Second, using a correlation technique, data registration for amplitude images can be performed. After registration, an interferogram image can be produced which reflects phase differences in two INSAR complex images. After phase unwarping and baseline estimation, a relationship between absolute phase and elevation is established. Finally, a DEM is generated.

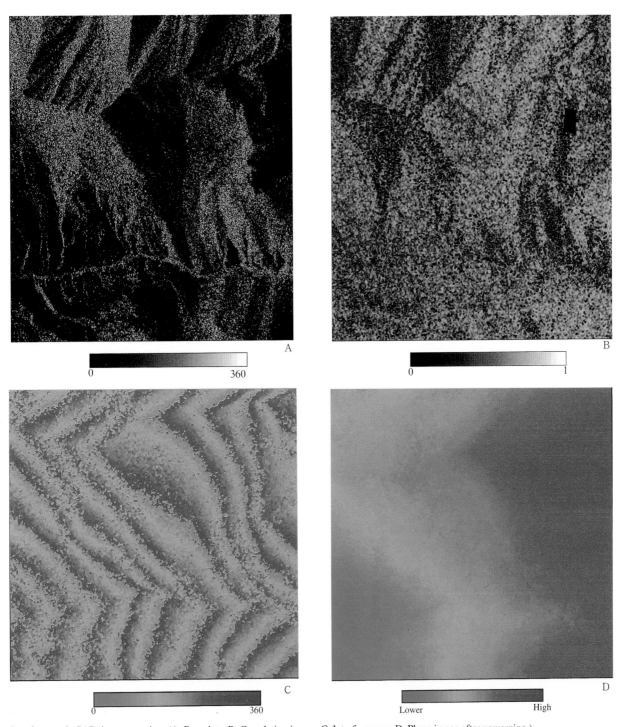

Interferometric SAR data processing. (A. Raw data, B. Correlation image, C. Interferogram, D. Phase image after unwarping.)

Extraction of Polarimetric Information

The C-band GlobeSAR polarimetric data was obtained on November 21, 1993 in the Zhaoqing area. In the false color image ($\sigma^{\circ}_{HV} + \sigma^{\circ}_{HV} + \sigma^{\circ}_{VV}$), there are several kinds of targets: Banana Trees (BA), Rice (RC), Harvested Rice (RD) with stems 20-30cm high, Water (WT), Buildings (BD), and water plant (WP).

These targets have different dielectric and geometrical features, which result in differences in polarization response graphs. BA, RC, and WT have the same feature of $\sigma^{\circ}_{HH} < \sigma^{\circ}_{VV}$ whereas RD has $\sigma^{\circ}_{HH} > \sigma^{\circ}_{VV}$. Using the differences in polarization information, such as the backscattering coefficient and degree of polarization, the targets were classified.

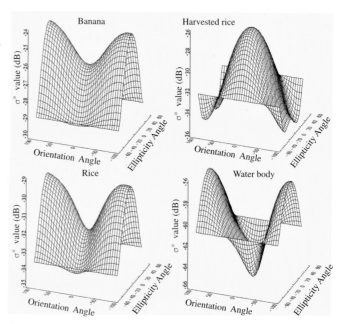

Polarization response graphs of banana tree, rice, harvested rice, and water body.

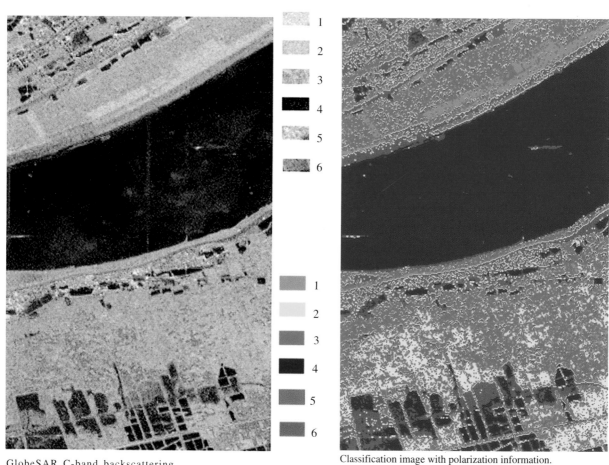

GlobeSAR C-band backscattering coefficients composite image of test area.

Classification image with polarization information. (using Neural network classifier input with σ_{HH}, σ_{HV}, σ_{VV}, H and V polarization degree information).

1. Banana   2. Rice   3. Harvested rice   4. Water body   5. Building   6. Aquatic plants

REFERENCES

Brewer, P.G., 1981, Oceanography: The Present and Future, Springer-Verlag, New York.

Beal R. C., P. S. DeLeonibus and I. Katz, 1981, Spaceborne Synthetic Aperture Radar for Oceanography, The Johns Hopkins University Press.

Campbell, F. H. A., Guo Huadong, Shao Yun and B. Brisco, 1996, *Proceedings of the Second Asia Regional GlobeSAR Workshop*, Science Press.

Chen Shupeng, 1986, Geoscience Analysis Atlas of Landsat Imagery in China, Science Press.

Chen Zhiming, 1986, Lake Retrogradation in Qinghai-Tibet Plateau and its Significance, *Marine and Limnology*, 17(3): 207 - 216.

Committee of Resources and Environment Remote Sensing in Loess Plateau of China, 1994, Forest map (1:500000) of Loess Plateau, Xian Cartographic Publishing House.

Deng Wanming, 1989, Cenozoic Volcanic Rocks in Northern Ali District of Tibet and Discussion of Intercontinental Subduction, *Journal of Petrology* (in Chinese), 5(3): 1 - 11.

Editorial Committee of 1:1 Million Land Use Map of China, 1987, Land Use Map of China, Science Press.

Elachi, C, 1988, Spaceborne Radar Remote Sensing: Applications and Techniques, IEEE Press.

Elachi, C., 1987, Introduction to the Physics and Techniques of Remote Sensing, John Wiley & Sons, Inc.

Evans, D. L., J. J. Plaut and E. Stofan, 1997, Overview of the Spaceborne Imaging Radar-C/X-band Synthetic Aperture Radar (SIR-C/X-SAR) Missions, *Remote Sensing of Environment*, 59(2):135-140.

Farr, T. G. and O. A. Chadwick, 1996, Geomorphic Processes and Remote Sensing Signatures of Alluvial Fans in the Kunlun Mountains, China, *Journal of Geophysical Research*, 101(E10): 23091-23100.

Fu, L. L., B. Holt, 1982, Seasat Views Oceans and Sea Ice With Synthetic-Aperture Radar, JPL Publication, 81-120.

Guo Huadong, 1991, Radar Image Analysis and Geologic Application, Science Press.

Guo Huadong, 1992, Airborne Synthetic Aperture Radar Experiments and Applications, China Science and Technology Press.

Guo Huadong, Xu Guanhua, 1995, Research on Spaceborne SAR Applications, China Science and Technology Press.

Guo Huadong, Wang Chao, Liao Jinjuan, Shao Yun and Wei Chenjie, 1995, Dual-frequency and Quad-polarization SAR Observations in Zhaoqing Region,China,*Geocarto International*, 109(3): 79-85.

Guo Huadong, Zheng Lizhong (eds.), 1996, Microwave Remote Sensing for Earth Observation, Science Press.

Guo Huadong, Zhu Liangpu, Shao Yun and Lu Xinqiao, 1996, Detection of Structural and Lithological Features Underneath Vegetation Canopy Using SIR-C/X-SAR Data in Zhao Qing Test Site of Southern China, *Journal of Geophysical Research*, 101(E10): 23101-23108.

Guo Huadong, 1997, Spaceborne Multifrequency, Polarimetric SAR and Interferometric SAR for Detecting Terrain Features, *Journal of Remote Sensing*, 1(1): 32-39.

Guo Huadong, Liao Jinjuan, Wang Changlin, Wang Chao, T. G. Farr and D. L. Evans, 1997, Use of Multifrequency, Multipolarization Shuttle Imaging Radar for Volcano Mapping in Kunlun Mountains of West China, *Remote Sensing of Environment*, 59(2).

Guo Huadong, Shao Yun, 1997, Airborne Dual Frequency and Full Polarization SAR Data Analysis, *Journal of Remote Sensing*, 1(2).

Guo Huadong, V. Singhroy and T. G. Farr (eds.), 1997, New Technology for Geosciences, VSP.

Guo Huadong, Wang Chao, Wang Xiangyun, Shi Yangshen, 1997, ERS-1 Scatterometer Data for Global Land Monitoring, *Journal of Remote Sensing*, 1(4).

Han Zhongshan, 1988, Landslides and Rockfalls of Yangtze Gorges, Geological Publishing House.

He Chunsun, 1986, Introduction to Taiwan Geology, Taiwan Geological Surveying Press.

Henderson, F. M. and A. J. Lewis , 1998, Manual of Remote Sensing: Principles and Applications of Imaging Radar, John Wiley and Sons, Inc.

Kennett, R. G. and F. K. Li, 1989, Seasat over-land scatterometer data, Part I: Global overview of the Ku-band backscatter coeficients, *IEEE Trans. Geosci. Remote Sensing*, 2: 592−605.

Institute of Hydrogeology and Engineering Geology of National Geological Bureau, 1971, Hydrogeology Atlas of the People's Republic of China, Cartographic

Publishing House.

Institute of Seismology and Institute of Seismo-geology, Bureau of National Seismology, 1982, Atlas of Typical Satellite Images for Active Structures in China, Seismic Publishing House.

Institute of Oceanic Forecasting, SOA, Institute of Oceanography Technology, SOA, Institute of Geography, Changchun, CAS, 1990, Airborne Remote Image for Sea Ice Atlas of Bohai Sea, Ocean Press, Beijing.

Li Bingyuan, Zhang Qingsong, Wang Fubao, 1991, Lake Evolutions in Kara-Kunlun and Kunlun, *Quaternary Research*, 1:64-71.

Li Shijie, Zheng Benxing, Jiao Keqin, 1993, Preliminary Discussion on Lakes in Western Kunlun Mountain, *Marine and Limnology*, 24(1): 37-44.

Li Wei Neng, 1985, Atlas of Geomorphology in China, Surveying and Mapping Publishing House.

Liu Dongsheng, 1991, Loess Plateau, Science Press.

Liu Gaohuan, 1996, Atlas of Sustainable Development in the Yellow River Delta, Surveying and Mapping Publishing House.

Long, M. W, 1983, Radar Reflectivity of Land and Sea, Artech House.

Madsen, S. N., H. A. Zebker, and J. A. Martin, 1993, Topographic Mapping Using Radar Interferometry: Processing Techniques, *IEEE Trans. Geosci. Remote Sensing*, 31(1): 246-256.

Meadows G. A. et al., 1983, Seasat Synthetic Aperture Radar Observations of Wave-Current and Wave-Topographic Interactions, *Journal of Geophysical Research*, 88 (C7): 4393-4406.

Shao Yun, Guo Huadong, Liu Hao, Li Junfei and Lu Xinqiao, 1995, Effect of Polarization and Frequency Using GlobeSAR Data on Vegetation Discrimination, *Geocarto International*, 109(3): 71-78.

Shao Yun, et al., 1995, The GlobeSAR Data for Vegetation Discrimination, Microwave Remote Sensing for Earth Observation, Science Press, 195-201.

Shao Yun, Guo Huadong, Liu Hao and Lu Xinqiao, 1996, GlobeSAR Data for Agriculture Applications-Potentials and Limitations, *Proceeding Second Asia Regional GlobeSAR Workshop*, Science Press, 79-83.

Shao Yun, F. Verjee and S. Staples, 1997, Cutting through cloud, *GIS Asia Pacific*, October/November, Singapore.

Shao Yun, Fan Xiangtao, Wang Cuizhen and Liu Hao, 1997, Estimation rice growth stage using Radarsat data, *Proceedings of IGARSS'97*, Vol.4, Singapore.

Shao Yun, et al. 1997, Effect of Frequency and Polarization on Target Detection, *Proceedings of IEAS/IWGIS'97*, Beijing.

Ulaby, F. T. and C. Elachi, 1990, Radar Polarimetry for Geoscience Applications, Artech House.

William Mcleish, et al., 1980, Synthetic Aperture Radar Imaging of Ocean Waves: Comparison with Wave Measuremets, *Journal of Geophysical Research*, 85:5003-5011.

Xu Guanhua, 1994, Theory and Application of Renewable Resources Remote Sensing for Three North Shelter Forest, China Forestry Publishing House.

Zhang Zhengmin, Li Hongjie, 1987, Atlas of the Yellow River Catchments, China Cartographic Publishing House.

DUE

WITHDRAWN